SPACE TELESCOPE SCIENCE INSTITUTE

T0233174

SYMPOSIUM SERIES: 19

*Series Editor* S. Michael Fall, Space Telescope Science Institute

# A DECADE OF EXTRASOLAR PLANETS AROUND NORMAL STARS

Humans have long thought that planetary systems similar to our own should exist around stars other than the Sun, yet the search for planets outside our Solar System has had a dismal history of discoveries that could not be confirmed. However, this all changed in 1995, with the past decade witnessing astonishing progress in this field; we now know of more than 200 extrasolar planets. These findings mark crucial milestones in the search for extraterrestrial life – arguably one of the most intriguing endeavors of modern science.

These proceedings from the Space Telescope Science Institute Symposium on extrasolar planets explore one of the hottest topics in astronomy today. Discussions include the *Kepler* mission, observational constraints on dust disk lifetimes and the implications for planet formation, and gravitational instabilities in protoplanetary disks. With review papers written by world experts in their fields, this is an important resource on extrasolar planets.

**SPACE TELESCOPE SCIENCE INSTITUTE**

Operated for NASA by AURA

# A decade of extrasolar planets around normal stars

## Proceedings of the Space Telescope Science Institute Symposium, held in Baltimore, Maryland May 2–5, 2005

*Edited by*
### MARIO LIVIO
*Space Telescope Science Institute, Baltimore, MD 21218, USA*

### KAILASH SAHU
*Space Telescope Science Institute, Baltimore, MD 21218, USA*

### JEFF VALENTI
*Space Telescope Science Institute, Baltimore, MD 21218, USA*

Published for the Space Telescope Science Institute

Operated for NASA by AURA

CAMBRIDGE
UNIVERSITY PRESS

CAMBRIDGE UNIVERSITY PRESS
Cambridge, New York, Melbourne, Madrid, Cape Town, Singapore,
São Paulo, Delhi, Dubai, Tokyo, Mexico City

Cambridge University Press
The Edinburgh Building, Cambridge CB2 8RU, UK

Published in the United States of America by Cambridge University Press, New York

www.cambridge.org
Information on this title: www.cambridge.org/9780521173308

First published 2008
First paperback edition 2011

*A catalogue record for this publication is available from the British Library*

ISBN 978-0-521-89784-6 Hardback
ISBN 978-0-521-17330-8 Paperback

# Contents

# Participants

| | |
|---|---|
| Adams, Elisabeth | Massachusetts Institute of Technology |
| Agol, Eric | University of Washington |
| Akinola, Femi | Corporate Insignia Consulting Limited |
| Alsubai, Khalid | St. Andrews University |
| Armitage, Philip | University of Colorado |
| Atcheson, Paul | Ball Aerospace |
| Batalha, Natalie | San Jose State University |
| Beckwith, Steven | Space Telescope Science Institute |
| Benedict, G. Fritz | McDonald Observatory |
| Bennett, David | University of Notre Dame |
| Borucki, William | NASA Ames Research Center |
| Boss, Alan | Carnegie Institution of Washington |
| Brown, Robert | Space Telescope Science Institute |
| Brown, Timothy | High Altitutde Observatory |
| Bullard, Floyd | Duke University |
| Burrows, Adam | University of Arizona |
| Caldwell, Douglas | SETI Institute |
| Calvet, Nuria | Smithsonian Astrophysical Observatory |
| Carpenter, John | California Institute of Technology |
| Clampin, Mark | NASA Goddard Space Flight Center |
| Clarkson, Will | Open University |
| Cox, Z. Nagin | Jet Propulsion Laboratory |
| D'Amario, James | Harford Community College |
| DeWitt, Curtis | University of Florida |
| Deming, Drake | NASA Goddard Space Flight Center |
| Dobbs-Dixon, Ian | University of California–Santa Cruz |
| Doering, Ryan | Space Telescope Science Institute |
| Durisen, Richard H. | Indiana University |
| Ebbets, Dennis | Ball Aerospace |
| Ehrenreich, David | Institut d'astrophysique de Paris |
| Elliot, James | Massachusetts Institute of Technology |
| Enoch, Becky | Open University |
| Fonda, Mark | NASA Ames Research Center |
| Fortney, Jonathan | NASA Ames Research Center |
| Frank, Adam | University of Rochester |
| Garside, Jeffrey | SETI Institute |
| Gilhooly, Jane | University of Maryland |
| Gilliland, Ronald | Space Telescope Science Institute |
| Godon, Patrick | Space Telescope Science Institute |
| Hales, Antonio | University College London |
| Harper-Clark, Elizabeth | Massachusetts Institute of Technology |
| Hartman, Joel | Harvard University |
| Hartnett, Kevin | NASA Goddard Space Flight Center |
| Haugabook, Sr., Ismail | Santa Monica College |
| Hauser, Michael | Space Telescope Science Institute |
| Hewagama, Tilak | University of Maryland |
| Hillenbrand, Lynne | California Institute of Technology |

| | |
|---|---|
| Hubickyj, Olenka | University of California/Lick Observatory and NASA Ames Research Center |
| Ivison, Rob | Royal Observatory Edinburgh |
| Jeletic, James | NASA Goddard Space Flight Center |
| Jenkins, Jon | SETI Institute and NASA Ames Research Center |
| Jensen, Eric | Swarthmore College |
| Kane, Julia | Massachusetts Institute of Technology |
| Kaye, Tom | Spectrashift Observatory |
| Kendrick, Stephen | Ball Aerospace & Technologies Corp. |
| Kerber, Florian | Space Telescope–European Coordinating Facility |
| Knutson, Heather | Harvard University |
| Koch, David | NASA Ames Research Center |
| Kochte, Mark | Computer Science Corporation |
| Kovacs, Geza | Konkoly Observatory |
| Krist, John | Jet Propulsion Laboratory |
| Lafreniere, David | University of Montréal |
| Latham, David | Harvard-Smithsonian Center for Astrophysics |
| Leckrone, David | NASA Goddard Space Flight Center |
| Leisawitz, David | NASA Goddard Space Flight Center |
| Levison, Harold | Southwest Research Institute |
| Li, Shulin | University of California–Santa Cruz |
| Lin, Douglas | University of California–Santa Cruz |
| Lissauer, Jack | NASA Ames Research Center |
| Livio, Mario | Space Telescope Science Institute |
| Lubow, Steve | Space Telescope Science Institute |
| Machalek, Pavel | The Johns Hopkins University |
| Malhotra, Renu | University of Arizona |
| Mamajek, Eric | Harvard-Smithsonian Center for Astrophysics |
| Marcy, Geoffrey | University of California–Berkeley |
| Margon, Bruce | Space Telescope Science Institute |
| Matthews, Jaymie | University of British Columbia |
| Mayor, Michel | Observatoire de Genève |
| McCullough, Peter | Space Telescope Science Institute |
| Meibom, Soren | University of Wisconsin–Madison |
| Meixner, Margaret | Space Telescope Science Institute |
| Meyer, Michael R. | University of Arizona |
| Miller, Joleen | University of Virginia |
| Moffat, Hope | Geo-Instruments |
| Murray, Norman | CITA, University of Toronto |
| Najita, Joan | National Optical Astronomy Observatory |
| Niedner, Malcolm | NASA Goddard Space Flight Center |
| Nota, Antonella | Space Telescope Science Institute |
| Olanitori, Olajide | Corporate Insignia Consulting Limited |
| Oliveira, Isabel | The Johns Hopkins University |
| Penny, Alan | Harvard-Smithsonian Center for Astrophysics |
| Pepper, Joshua | Ohio State University |
| Pilcher, Carl | NASA Headquarters |
| Pont, Frédéric | Observatoire de Genéve |
| Pringle, James | Institute of Astronomy |
| Quintana, Elisa V. | NASA Ames Research Center |

| | |
|---|---|
| Reid, I. Neill | Space Telescope Science Institute |
| Richardson, Jeremy | National Research Council and |
| | NASA Goddard Space Flight Center |
| Robberto, Massimo | Space Telescope Science Institute |
| Sahu, Kailash | Space Telescope Science Institute |
| Sari, Re'em | California Institute of Technology |
| Sasselov, Dimitar | Harvard-Smithsonian Center for Astrophysics |
| Schneider, Glenn | Steward Observatory, Universty of Arizona |
| Schultz, Alfred | Space Telescope Science Institute |
| Seager, Sara | Carnegie Institution of Washington |
| Shen, Zhixia | Peking University |
| Soderblom, David | Space Telescope Science Institute |
| Sparks, William | Space Telescope Science Institute |
| Stauffer, John | Spitzer Science Center, |
| | California Institute of Technology |
| Valenti, Jeff | Space Telescope Science Institute |
| Varniere, Peggy | University of Rochester |
| Vidal-Madjar, Alfred | Institut d'Astrophysique de Paris, CNRS, |
| | Université Pierre et Marie Curie |
| Werner, Michael | Jet Propulsion Laboratory |
| Whitmore, Brad | Space Telescope Science Institute |
| Wisdom, Jack | Massachusetts Institute of Technology |
| Wiseman, Jennifer | NASA Headquarters |
| Yusuf, Affez Olawale | Ibramed Nigeria Limited |

# Preface

The Space Telescope Science Institute Symposium on *A Decade of Extrasolar Planets around Normal Stars* took place during 2–5 May 2005.

These proceedings represent only a part of the invited talks that were presented at the symposium. We thank the contributing authors for preparing their manuscripts.

The past decade has witnessed astonishing progress in this field. While before 1995 not a single planet around a normal star was known outside the Solar System, we now know of more than 160 such planets. Furthermore, while the initial discoveries were of Jupiter-size planets, currently we know of an extrasolar planet with a mass of about five Earth masses. These findings mark crucial milestones in the search for extraterrestrial life—arguably one of the most intriguing endeavors of science today.

We thank Sharon Toolan of ST ScI for her help in preparing this volume for publication.

Mario Livio
Kailash Sahu
Jeff Valenti
*Space Telescope Science Institute*
*Baltimore, Maryland*

# Extrasolar planets: Past, present, and future

## By ALAN P. BOSS

Department of Terrestrial Magnetism, Carnegie Institution, Washington, DC 20015, USA

Human beings have long thought that planetary systems similar to our own should exist around stars other than the Sun. However, the astronomical search for planets outside our Solar System has had a dismal history of decades of discoveries that were announced, but could not be confirmed. All that changed in 1995, when we entered the era of the discovery of extrasolar planetary systems orbiting main-sequence stars. To date, well over 130 planets have been found outside our Solar System, ranging from the fairly familiar to the weirdly unexpected. Nearly all of the new planets discovered to date appear to be gas giant planets similar to our Jupiter and Saturn, though with very different orbits about their host stars. In the last year, three planets with much lower masses have been found, similar to those of Uranus and Neptune, but it is not yet clear if they are also ice giant planets, or perhaps rock giant planets, i.e., super-Earths. The long-term goal is to discover and characterize nearby Earth-like, habitable planets. A visionary array of space-based telescopes has been planned that will carry out this incredible search over the next several decades.

## 1. Introduction

Natural philosophers had hypothesized centuries ago that other planetary systems orbited the many stars in the night sky, that the Solar System was not unique. Up until 1995, however, there was no reproducible astronomical evidence to support this visionary viewpoint (Boss 1998). One decade later, a new field of astronomy has been born, with rapid observational progress that threatens to far outpace theoretical efforts to keep up. Major discoveries continue to appear on roughly a monthly basis, an unprecedented level of advancement in any field of science. While we have yet to find a true Solar System analogue, the planetary systems discovered so far leave little doubt that we will soon be discovering planetary systems that will be hospitable to the existence of Earth-like planets, the ultimate goal of this entire field of research.

Surprisingly, the very first planetary-mass bodies discovered outside our Solar System were roughly Earth-mass planets orbiting the pulsar PSR 1257+12 (Wolszczan & Frail 1992). These objects must have formed after their host star underwent a supernova explosion, as the explosion and accompanying stellar-mass loss would likely have removed any pre-existing planets. The fact that Earth-mass bodies later managed to form in a disk around the neutron star was taken as a strong proof of the resiliency of the planet-forming process of collisional accumulation of solids into larger bodies, even in a hostile environment. Nevertheless, the paucity of evidence for planetary systems around normal, main-sequence stars caused significant concern even after the pulsar planets were announced in 1992—where were the gas giant planets?

## 2. Past

The fantastic success of the *Hubble Space Telescope* (*HST*) at a wide range of astronomical observations has spoiled us all—we have come to expect to see images on a frequent basis of gorgeous cosmic locales, from nearby comets to distant galaxies. Taking an image of a nearby extrasolar planet is no more difficult than taking an image of a distant galaxy ($V \sim 30$ mag), except for the fact that the faint planet is located right

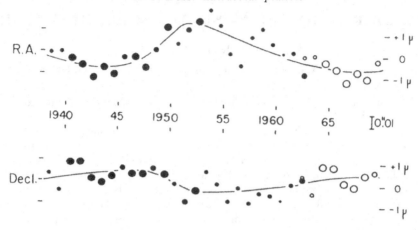

FIGURE 1. Apparent astrometric detection of a 1.6 Jupiter-mass planet orbiting Barnard's Star on a 24-year period, eccentric ($e = 0.6$) orbit (solid line), based on yearly means of Sproul Observatory data (van de Kamp 1963).

next to its much brighter host star. At optical wavelengths, the host star may outshine its planets by a factor of $10^9$ or more. *HST* was not designed to be able to separate out the light of a planet from its host star.

As a result, essentially all of our information about extrasolar planets has come from indirect detection methods, where the existence of the planet is inferred based on the gravitational reflex motion of its host star as it orbits the center of mass of the entire system. This motion can be detected in several different ways, with the first studies having sought the back-and-forth motion of the star on the plane of the sky with respect to nearly stationary, background stars: the astrometric technique.

### 2.1. *Barnard's Star*

In 1937 Peter van de Kamp became the Director of Swarthmore College's Sproul Observatory. In the next year, he began a long term astrometric program to search for low-mass companions to nearby stars with the Observatory's 24-inch refractor telescope. One of the first stars he added to the target list was Barnard's Star, discovered in 1916 by E. E. Barnard. Barnard's Star is a red dwarf with a mass of ~0.15 $M_\odot$, close enough at 1.8 pc that only the Alpha Centauri triple system is closer to the Sun, but so faint that it cannot be seen with the naked eye. As a low-mass star very close to the Sun, it is an excellent candidate for an astrometric planet search: a Jupiter-like planet would force Barnard's Star to wobble over a total angle of ~0.04 arcsec, a wobble of several microns on the photographic emulsions used to record the trajectory of Barnard's Star, as it sped across the sky at 10 arcsec per year.

Astrometric claims for very low-mass companions to the stars 70 Ophiuchi and 61 Cygni had been made in 1943 by two different groups, but these claimed detections of objects with masses in the range of 10 to 16 Jupiter masses could not be verified and were soon discarded, if not forgotten. Only after taking 2,400 observations of Barnard's Star for over two decades was van de Kamp ready to announce his discovery of a planet, given this unfortunate prior history. In 1963 van de Kamp announced that he had found the first extrasolar planet: a planet with a mass only 60% greater than that of Jupiter, orbiting Barnard's Star with a period of 24 years (van de Kamp 1963). In order to be certain of its reality, he had waited for an entire orbital period to elapse before making the an-

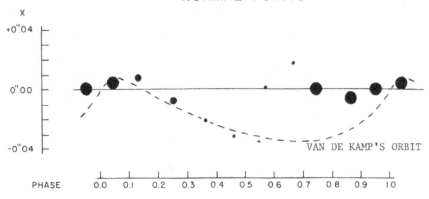

FIGURE 2. Refutation of a planetary companion to Barnard's Star, based on an independent data set and astrometric analysis (Gatewood & Eichhorn 1973). The size of the data points indicates their weight in the astrometric solution—smaller data points have larger errors. Only one astrometric axis ($x$) is shown.

nouncement (Figure 1). The semimajor axis of the planet was 4.4 AU, similar to Jupiter's 5.2 AU, and the fact that the orbit was considerably more eccentric than Jupiter's orbit did not raise too many questions about what had been found. Astronomers expected gas giant planets to exist elsewhere, and within a few years, Barnard's Star literally became the textbook example of another star with a planetary system.

On the advice of a senior professor at the University of Pittsburgh who must have doubted the existence of van de Kamp's planet, George Gatewood undertook a second study of Barnard's Star. Investigating an independent collection of photographic images of Barnard's Star, Gatewood used a new plate-measuring engine at the U.S. Naval Observatory to remove the human element from the plate-measuring process. Applying mathematical techniques derived by his thesis advisor, Heinrich Eichhorn, Gatewood analyzed 241 plates taken at the University of Pittsburgh's Allegheny Observatory and at the van Vleck Observatory in Connecticut. Surprisingly, their analysis was not able to confirm the existence of a planet orbiting Barnard's Star (Gatewood & Eichhorn 1973). Figure 2 shows that their data precluded the large amplitude wobble in van de Kamp's data.

The situation got much worse for van de Kamp's planet in the same year, when a colleague of his at the Sproul Observatory published an analysis of the astrometry of GL 793, another nearby red dwarf star. John Hershey had found that both GL 793 and Barnard's Star were wobbling about their proper motion across the sky in much the same way. When one star zigged, so did the other. When one star zagged, so did the other. This could only mean one thing: systematic errors in the Sproul refractor. In retrospect, the spurious zig-zags could be traced to several changes that had been made to the optical system, but evidently had not been completely corrected for in the error terms for the astrometric solutions: a new cast iron cell for the 24-inch lens and new photographic emulsions in 1949, and a lens adjustment in 1957.

After 1973, the strong evidence for Barnard's Star having a gas giant planet began to fade from view and from the textbooks. Astronomers have continued to monitor Barnard's Star for planetary-induced wobbles, as it remains an attractive target. Van de Kamp also continued to pursue Barnard's Star, convinced that some day a planet

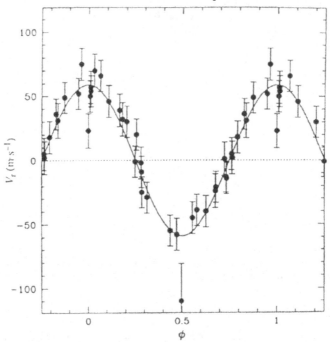

FIGURE 3. Discovery data for the first extrasolar planet around a main sequence star, 51 Pegasi (Mayor & Queloz 1995), detected by measuring the periodic Doppler velocity shift of 51 Pegasi induced by the planetary companion. The solid line shows the Doppler shift expected for a solar-mass star orbited by a 0.6 Jupiter-mass planet on a circular orbit with a semimajor axis of 0.05 AU.

would be found in its grasp. Van de Kamp died in 1995, just before the first reproducible evidence for an extrasolar giant planet was announced to the world.

## 2.2. *51 Pegasi*

The existence of an extrasolar planet can be inferred indirectly, not only by measuring the two-dimensional, astrometric wobble in the plane of the sky, but also by searching for the one-dimensional oscillation of the star to-and-fro along the line-of-sight to the star by measuring the Doppler shift of the star's spectral lines. For a Jupiter-like planet orbiting a solar-type star, the Doppler shift has a semiamplitude of $13$ m s$^{-1}$. Beginning in the late 1970s, a group at the University of British Columbia had pioneered a technique of using a cell of hydrogen fluoride gas in the optical path of the telescope as a source of stable reference lines, which could be used to measure tiny Doppler shifts in stellar spectra. By 1995, however, they had searched for over a decade and had found no unambiguous evidence for planets in the two dozen stars they had been following—their results appeared to place upper limits only on the masses of planets that might exist in orbit around their target stars (Walker et al. 1995). It looked like extrasolar Jupiters might not be very common, contrary to long-standing theoretical and philosophical expectations.

Shortly after the Walker et al. (1995) results appeared, a new claim for the detection of a gas giant planet was announced by a team composed of Michel Mayor and Didier Queloz of the Geneva Observatory. Duquennoy & Mayor (1991) had published the definitive catalog of binary stars in the solar neighborhood, including binaries found by several different methods, but primarily by searching for the Doppler spectroscopic wobble. As a result of this survey, Mayor had a list of roughly 200 nearby solar-type stars that appeared to be single, though a number of them showed evidence for having a very low-mass companion. In 1994, a new spectrometer on the 2 m telescope at the Haute Provence Observatory with a spectral accuracy of ~13 m s$^{-1}$ permitted Mayor to begin a serious spectroscopic search for extrasolar giant planets. By the summer of 1995, Mayor & Queloz had struck oil—the solar-type star 51 Pegasi appeared to be wobbling with an amplitude well above their measurement errors. Following a final, confirmatory observing run in September 1995, Mayor & Queloz (1995) announced their discovery of a ~0.6 Jupiter-mass planet on a *circular* orbit (Figure 3), as expected based on Jupiter's low eccentricity ($e \sim 0.05$) orbit. While the mass and circular orbit of the planet seemed normal, its orbital period was far from the expectations for an extrasolar Jupiter: 4.23 days rather than 12 or so years. This meant that the planet was orbiting its star 100 times closer than Jupiter orbits the sun, at a distance of ~0.05 AU rather than out at 5.2 AU. At that distance from its star, 51 Pegasi's planet would have an atmosphere that was thermally evaporating: it would be a "hot Jupiter."

Was 51 Pegasi's planet real or not? Following the announcement in October 1995, several other groups of astronomers ran to their telescopes to see if they could confirm this audacious claim—and they did. The first team to confirm the reality of 51 Pegasi's planet was that of Geoff Marcy and Paul Butler, using the Shane reflector on Mount Hamilton (Marcy et al. 1997). Several other teams followed in their footsteps and also confirmed the detection of 51 Pegasi's planet. Suddenly the idea of extrasolar planets around solar-type stars did not seem so preposterous anymore.

Marcy and Butler had begun their spectroscopic planet search in the late 1980s, using an iodine cell as a reference source for their spectrometer (since hydrogen fluoride is a deadly gas), and they were soon achieving accuracies of 10 m s$^{-1}$ or better. Given their significant head start on the Swiss group, Marcy and Butler began a frantic effort to reduce their many years of data, allowing them to announce the discovery of two more planets in January, 1996—planets orbiting the solar-type stars 47 Ursae Majoris (Butler & Marcy 1996) and 70 Virginis (Marcy & Butler 1996). The field of extrasolar planets had truly been born. *Time Magazine* celebrated the event with a cover story breathlessly entitled, "Is Anybody Out There? How the discovery of two planets brings us closer to solving the most profound mystery in the cosmos."

## 3. Present

In a field as fast moving as extrasolar planets, any attempt to summarize the current status is doomed to appear rather antique in a short period of time. Several major discoveries have been announced in the two months between the May 2005 Symposium at ST ScI and the writing of this summary (early July 2005), with more sure to follow once the Symposium proceedings go to press. With that caveat, the status of the field as of the time of the May 2005 Symposium can be summarized by the plot of discovery space shown in Figure 4. Here, the masses of extrasolar planets are shown as a function of the semimajor axes of their orbits. By May 2005, well over 130 planets had been discovered and submitted for publication in refereed journals. The International Astronomical Union's Working Group on Extrasolar Planets maintains a list of extrasolar planets that meet the Group's requirements for inclusion on their web site at http://www.dtm.ciw.edu/boss/iauindex.html. This web site also addresses the question of defining what is and what is not a "planet."

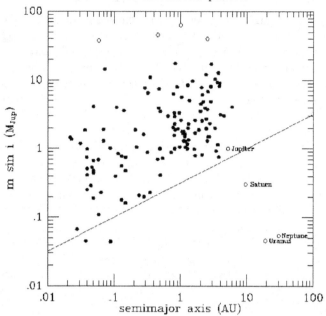

FIGURE 4. Extrasolar planet discovery space as of May 2005, showing primarily minimum masses of planets discovered by Doppler spectroscopy as a function of orbital semimajor axis. The oblique dashed line shows how the sensitivity limit of Doppler spectroscopy depends on semi-major axis for accuracies of $\sim$2–3 m s$^{-1}$ and a signal-to-noise ratio of $\sim$4—short-period, massive planets are the easiest to detect. Brown dwarfs (open symbols near the top of discovery space) are infrequent companions to solar-type stars.

### 3.1. *Doppler Spectroscopy*

Nearly all of the planets shown in Figure 4 were discovered by the Doppler spectroscopy method, which yields only a lower limit on the mass of the planet because of the unknown orientation of the planet's orbit with respect to the line-of-sight to the star. If the planetary orbit is being observed nearly pole-on, then the mass of the companion is larger by a factor of $1/\sin i$ (where $i$ is the inclination of the planet's orbit; $i = 0$ for pole-on) than the minimum mass plotted in Figure 4. Assuming that planetary orbital planes are randomly distributed, the typical true planetary mass should be a factor of $4/\pi$ larger than the minimum mass, i.e., $\sim$1.3 times higher.

Given the drive to find Solar System analogues and their spectroscopic suitability, most of the target stars in the ground-based spectroscopic planet searches have been late F, G, and K dwarfs, though these searches have also been extended to early M dwarfs. It is evident from Figure 4 that solar-type stars tend not to have brown dwarf companions, i.e., companions capable of burning deuterium (requiring a mass greater than $\sim$13 Jupiter masses for solar composition), but not hydrogen (implying a mass less than $\sim$75 Jupiter masses). This is consistent with the lack of binary companions with mass ratios much larger than 10:1 (Duquennoy & Mayor 1991)—solar-mass primaries generally do not have brown dwarf companions.

Figure 4 also makes it clear that such stars do typically have planetary-mass companions, though evidently the process that produces these objects does not always adhere to the IAU definition of a planet as being an object less massive than 13 Jupiter masses. Note that the absence of planets with semimajor axes greater than a few AU does not imply their nonexistence, but rather is a result of the need to follow a target star for an entire planetary orbit before announcing a detection in order to minimize the risk of interpreting noisy data as a detection. The spectroscopic surveys are just now entering the phase of their programs where they have been monitoring stars with sufficient accuracy (a few m s$^{-1}$) long enough (a decade or so) to begin to detect long-period planets similar to Jupiter.

Figure 4 shows that the range of masses of extrasolar planets is considerably greater than in our Solar System: planets with masses ten times that of Jupiter exist, as well as masses smaller than that of Saturn. Recently, the attainment of Doppler spectroscopy precisions of ~1 m s$^{-1}$ means that the lower mass limit has been extended down to Neptune masses by the discoveries of planets orbiting GJ 436 (Butler et al. 2004), Mu Arae (Santos et al. 2004), and $\rho^1$ Cancri (McArthur et al. 2004). These three might well represent the first examples of a new class of planets, i.e., they could be ice-giant planets, given the similarity of their masses to those of Uranus and Neptune, or they might be super-Earths, rocky planets with masses well above that of Earth and Venus. The former explanation seems to be inconsistent with the occurrence of several gas-giant planets on longer-period orbits in both the Mu Arae and $\rho^1$ Cancri systems, implying that the Neptune-mass planets formed inside the orbital radii of their gas giants, and then migrated inward. While this explanation seems most plausible, it will remain for a transit detection of a "hot Neptune" to measure one of these planets' mean densities, and so determine whether it is composed primarily of rock or of rock and ice/water.

The orbital radii evident in Figure 4 cover a wide range—from the semimajor axes of 0.02 AU of the "hot Jupiters" out to 5.2 AU for the "cold Jupiters," with a number of "warm Jupiters" orbiting in between. The close-in orbits of the hot and warm Jupiters imply significant post-formational inward orbital migration, given the difficulties in forming gas giants so close to their stars. Perhaps most surprisingly, many of the orbits are highly eccentric, making the low eccentricities of the major planets in our Solar System seem out of the ordinary, rather than the norm. The origin of these eccentricities has become another major theoretical puzzle—are they a result of the formation process or of the orbital-migration process?

It is notable that there have been no discoveries of extrasolar planets to date with the astrometric technique, though there have been two astrometric measurements of previously-known planets (for GJ 876 and $\rho^1$ Cancri) using the Fine Guidance Sensors of *HST* (McArthur et al. 2004).

## 3.2. *Transits*

The first planet seen to transit its host star was the hot Jupiter orbiting HD 209458, detected by Doppler spectroscopy (Charbonneau et al. 2000; Henry et al. 2000). Because of the short-period orbits of the hot Jupiters, orbiting at roughly 10 stellar radii, the chances of having the orbit of a hot Jupiter aligned so as to lead to a transit are roughly 10%. We had to wait for the tenth hot Jupiter to be discovered by spectroscopy before one of them was discovered to be a transiting planet—hopefully we will be luckier with the hot Neptunes.

HD 209458's planets provided the first strong evidence that many, if not most, of the objects in Figure 4 are indeed gas-giant planets. A transit fixes the orbital inclination, and thus the mass of the planet, and the depth of the transit allows the radius of the planet to be determined as a fraction of its host star's radius. HD 209458's planet's mass is $\approx 0.7$ $M_J$, and it has a radius and a density roughly equal to that expected for a hot Jupiter. In addition, sodium was detected in its atmosphere (Charbonneau et al. 2002), as predicted for a hot Jupiter (Seager & Sasselov 2000).

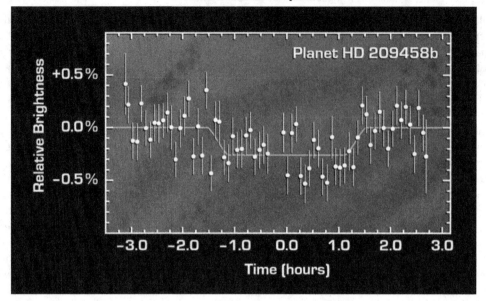

FIGURE 5. *Spitzer Space Telescope (SST)* 24-micron photometry of the star-planet system HD 209458 during a secondary eclipse of the planet by the star, constituting the first direct detection of light from an extrasolar planet (Deming et al. 2005). A similar detection was accomplished for the TrES-1 system by Charbonneau et al. (2005) using *SST*.

While Doppler spectroscopy has been by far the leader at detecting new planets, the transit detection technique has now accounted for the discovery of six, all confirmed by follow-up Doppler spectroscopy. The first planet detected by transit photometry was a hot Jupiter orbiting a star toward the galactic bulge (Konacki et al. 2003), that had been observed to have photometric variations consistent with a transiting planet (Udalski et al. 2002). This planet is known by the name of the transiting event, OGLE-TR-56. The Optical Gravitational Lensing Experiment (OGLE) project (Udalski et al. 2002) at the Las Campanas Observatory in Chile has discovered a large number of possible planetary transits, and four more so far have turned out to be caused by planets, all as a side benefit of the OGLE project. A sixth transiting planet (TrES-1) has been found by a new ground-based transit search program, the Transatlantic Exoplanet Survey (Alonso et al. 2004). Because transit surveys preferentially find short-period planets, all of the planets found to date by transits are hot Jupiters, though they are mostly smaller and denser than HD 209458's planet (Sozzetti et al. 2004).

### 3.3. *Microlensing*

A third technique that has found a planet around a main-sequence star is microlensing, where the photometric variations caused by gravitational bending of background starlight by a foreground star can be enhanced for a period of a few days by a planet orbiting at the Einstein radius. The first microlensing detection was accomplished by Bond et al. (2004) associated with the microlensing event known as OGLE 2003-BLG-235/MOA 2003-BLG-53. [Clearly there is a need for more succinct names for some of these extrasolar planets.] The inferred planet has a mass of ∼1.5 Jupiter masses and orbits at ∼3 AU from the presumed main sequence host star.

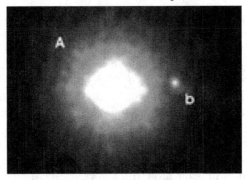

FIGURE 6. The apparent first spatially-resolved image of an extrasolar planet (b) orbiting ~100 AU from the young star GQ Lup (A). The image was obtained with the adaptive optics instrument NACO on the VLT by Neuhäuser et al. (2005).

### 3.4. *Pulsar Timing*

Astronomers have continued to search for more planetary-mass companions to pulsars by looking for minute variations in the pulsars' precise pulsation periods, as was used by Wolszczan & Frail (1992) to discover the pulsar PSR 1257+12 planetary system. However, searches of over 100 pulsars to date have yielded very little evidence for more planetary-mass companions, unlike main-sequence stars. The exception is the detection of a gas-giant planet-mass companion to a binary star system containing a white dwarf and the pulsar PSR B 1620-26 in the M4 globular cluster by Sigurdsson et al. (2003). Evidently gas-giant planets can form in regions quite different from the galactic disk, including extremely metal-poor environments such as an ancient globular cluster.

### 3.5. *Direct Detections*

The field of extrasolar planets took another enormous leap forward in 2005 with the announcement of several direct detections of extrasolar planets. The previous evidence for sodium in the atmosphere of HD 209458's planet (Charbonneau et al. 2002) was obtained by noting a depletion of the host star's light at the wavelengths of the sodium doublet lines during planetary transits, so formally speaking, this discovery did not detect photons from the planet itself, but rather the absence of stellar photons that had been absorbed in the upper atmosphere of the planet. All that changed in 2005, when the *Spitzer Space Telescope* (*SST*) enabled the first direct detection of the light from two transiting planets, HD 209458 (Figure 5; Deming et al. 2005) and TrES-1 (Charbonneau et al. 2005). Because the hot planet is relatively bright at mid-infrared wavelengths, when the planet disappears behind the star (the secondary eclipse) the total amount of mid-infrared light is observed to decrease by a measurable amount. When observed out of eclipse, then, the extra photons must be coming from the planet.

While pathbreaking, the *SST* direct detections of the HD 209458 and TrES-1 hot Jupiters were not direct detections in the sense of spatially resolving the light from the planet from that of the star. That honor appears to have been reserved for the detection of a multiple-Jupiter-mass planet in orbit around the T Tauri star GQ Lup. Neuhäuser et al. (2005) were able to obtain an image of GQ Lup's planet (Figure 6) by using the adaptive optics system on one of the Very Large Telescopes (VLT) in Chile. While the mass of the planet is uncertain, it could exceed the 13-Jupiter-mass upper bound for being a planet. The young age of this system (~1 Myr), when combined with recent models of the evolution of newly formed gas-giant planets, implies that this is indeed the first detection of a spatially-resolved extrasolar planet, though at a puzzlingly large orbital separation of ∼100 AU. This discovery has produced yet another challenge for the theorists: forming a gas-giant planet at such a large distance is likely to be difficult, implying that GQ Lup's planet may have been kicked outward to its present locale.

Prior to the discovery of GQ Lup's planetary candidate, another excellent discovery was made by the VLT. Chauvin et al. (2004) imaged a roughly 5-Jupiter-mass companion to the brown dwarf 2M1207. Because 2M1207 has a mass itself of only about 25 Jupiter masses, this system has a mass ratio of ∼5:1, typical of binary star systems. The two components of the 2M1207 system thus seem most likely to have formed simultaneously during the collapse and fragmentation process that forms binary and multiple star systems. While the very low mass of the companion seems to place it in the planet category, the fact that it is in orbit around a brown dwarf argues against awarding this discovery the prize of the first spatially resolved, direct detection of a "planet." A number of planetary-mass objects had previously been imaged in regions of recent star formation, with masses as low as that of 2M1207's secondary, and these objects are believed to have been formed directly by the same star formation process that leads to main-sequence stars and brown dwarfs. Accordingly, such objects are perhaps better referred to as "sub-brown dwarfs."

## 4. Future

It is hard to think of an area of astronomy that has a more exciting or promising future than that of extrasolar planets. In ten years, a new field has been created—with ongoing, frequent, major discoveries—and the pace continues to quicken as more astronomers shift their research interests in this direction. While much has already been learned, so much more remains to be discovered that it boggles the mind. We have learned about some of the properties of a bit more than 100 planets out of the billions of planets that appear to exist in our galaxy alone. The variety of extrasolar planets discovered to date—and those remaining to be found—may approach the variations observed in stellar populations and galaxy types. Ground-based observatories plan to continue to lead the way forward. Both the U.S. National Aeronautics and Space Administration (NASA) and the European Space Agency (ESA) have recognized the importance of this new field of endeavor, and have made far-reaching plans to build a series of ambitious space telescopes that will carry the search for extrasolar planets to its ultimate goal—Earth-like planets capable of originating and sustaining life.

### 4.1. *Ground*

Ground-based observatories bear the brunt of current extrasolar planet search activities, ranging from the Doppler spectroscopy programs of the Geneva Observatory, California/Carnegie, University of Texas, and other groups, to the several dozen transit and microlensing search programs underway around the world. The upgraded CCDs on the HIRES spectrograph on the Keck I telescope in Hawaii, the UVES spectrograph on the VLT, and the HARPS spectrometer on the 3.6 m ESO telescope at La Silla have allowed astronomers to push their Doppler precision to higher and higher levels, to the point where residual velocity jitters as small as ∼1 m s$^{-1}$ are now achievable, comparable to the convective velocity fluctuations in the photospheres of chromospherically quiet target stars. Maintaining these levels of Doppler precision for the next decade will allow the ground-based Doppler surveys to detect planets with masses similar to that of Saturn orbiting at 5 AU, and to detect planets with masses well below Neptune's mass on short period orbits—"hot Earths" could be found.

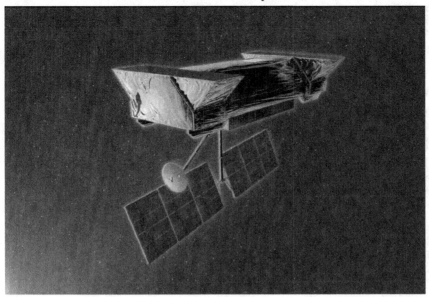

FIGURE 7. NASA's *Space Interferometry Mission (SIM)* is planned for launch around 2011. *SIM* will be an optical interferometer with a baseline of 9 m, allowing it to detect Earth-like planets orbiting the closest stars.

While NASA's plans for beginning an astrometric planet search with the Keck Interferometer have been delayed by concerns about installing the four 1.8 m Keck Outrigger Telescopes on Mauna Kea, the VLT's plan to combine four 1.8 m auxiliary telescopes with the four VLT 8.2 m Unit Telescopes on Paranal into the VLT Interferometer is well underway. The four VLTI auxiliary telescopes are expected to begin science operations by 2007. The aim is to attain astrometric accuracies of order 20 microarcsec, sufficient to detect Neptune-mass planets on long-period orbits.

Ground-based transit surveys are limited by the Earth's atmosphere to detecting planets with physical radii similar to that of Jupiter in orbit about solar-radius stars. Reaching down to smaller radius planets will require space-based telescopes.

### 4.2. *Space*

The Canadian *Microvariability and Oscillations of Stars (MOST)* microsatellite has initiated the era of space-based transit photometry telescopes. While *MOST* is limited by its modest aperture (0.15 m) to photometric monitoring of fairly bright stars, it does offer an improvement by a factor of $\sim$10 in photometric accuracy compared to the ground. CNES/ESA's Convection, Rotation, and planetary Transits (Corot) mission is scheduled for launch in 2006. Like *MOST*, Corot's first priority will be asteroseismology of a relatively small number of stars, but it will also be able to search for transits by planets as small as Neptune-mass on short-period orbits. Corot's 0.3 m telescope will spend roughly five months staring at the same region of the sky, allowing planets with orbital periods of a month or so to be reliably detected.

Corot will be followed in 2008 by the launch of NASA's *Kepler* Mission, a 0.95 m Schmidt telescope that is specifically designed to detect Earth-like planets. *Kepler* will stare at a field 105 square degrees in size in Cygnus containing several hundred thousand stars, and will pick out the best $\sim$10$^5$ stars for continuous monitoring during the four-year-long prime mission. *Kepler*'s CCDs have a photometric precision sufficiently high to allow detection of an Earth-mass planet by the dimming of a star's light by 0.01%.

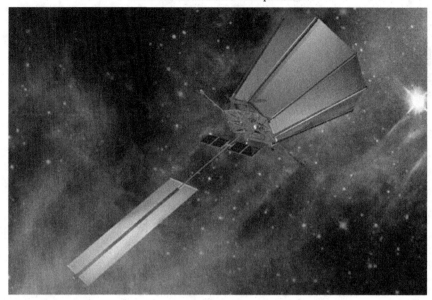

FIGURE 8. NASA's *Terrestrial Planet Finder Coronagraph* (*TPF-C*), planned for launch after 2016, will be a 3.5 m × 8 m optical telescope with a coronagraphic design, allowing Earth-like planets to be imaged next to their host stars.

Once the false alarms caused by situations such as unresolved background spectroscopic binaries blended with foreground stars are eliminated by follow-up observations, *Kepler* should be able to discover several dozen Earth-mass planets orbiting in the habitable zones of their stars. *Kepler* will be the first mission to determine the frequency of Earth-like planets, a key factor in planning for the next steps, or at least to place an upper limit on this frequency. ESA's *Gaia* Mission will be similar to *Hipparcos*: an astrometric survey of the entire sky, but with sufficient mission-end astrometric accuracy to detect giant planets around thousands of stars. *Gaia* is intended for launch in 2010 on a five-year mission.

NASA plans to launch the *Space Interferometry Mission* (*SIM*) around 2011 (Figure 7). The 9 m baseline for *SIM*'s optical interferometer will allow it to achieve a single measurement astrometric accuracy of ∼1 microarcsec. With repeated observations, *SIM*'s accuracy will allow the detection and determination of the orbits of planets as low in mass as Earth orbiting the closest stars. Being an astrometric telescope, *SIM* will determine the true masses of the planets it finds, not just lower limits. *SIM* will help to survey the solar neighborhood for potential targets for subsequent space missions intended to characterize Earth-like planets.

Following *SIM*, the next major step in NASA's Navigator Program is to launch the first of two *Terrestrial Planet Finders*. The first *TPF* mission will be an optical coronagraph (*TPF-C*), intended for launch around 2016 (Figure 8). *TPF-C* will have a 3.5 m × 8 m monolithic primary mirror combined with a series of coronagraphic optical elements and adaptive optics deformable mirrors that will strip away the star's direct, diffracted, and scattered light, allowing planets as faint as the Earth to be imaged on habitable zone orbits around nearby stars. *TPF-C* will not only seek to detect Earth-like planets beyond those discovered by *SIM*, but it will also have a low-resolution spectrographic capability that will allow it to search for atmospheric signatures of a habitable planet—such as carbon dioxide, water, and ozone.

FIGURE 9. NASA's *Terrestrial Planet Finder Interferometer* (*TPF-I*) is planned for launch after 2020, consisting of four collector spacecraft operating at mid-infrared wavelengths and a combiner spacecraft. ESA has a similar plan for a space telescope named *Darwin*. A joint NASA/ESA mission is being considered.

Figure 9 shows an image of NASA's *Terrestrial Planet Finder Interferometer* (*TPF-I*), the successor to *TPF-C*. *TPF-I* is planned for launch after 2020, and will be a free-flyer with four collector spacecraft and a single beam combiner spacecraft. The collector telescopes will have apertures of ∼3.5 m and will operate at mid-infrared wavelengths. With a ∼100 m baseline, *TPF-I* will be able to search for habitable Earths to a greater distance than *TPF-C*, allowing a larger number of stars to be searched. Once the extrasolar Earths are detected, *TPF-I* will use its spectroscopic capabilites to search for biosignatures such as carbon dioxide, water, oxygen, and methane. If the last two molecules can be found in the atmosphere of the same planet, that would be a strong indication that the planet is not only inhabitable, but may be even be inhabited, as methane and oxygen would soon combine chemically in the absence of robust production mechanisms for them. Abiotic chemistry may not be able to produce oxygen and methane in abundance and coexistence—some form of life seems to be required, at least based on what we know about Earth. Finding evidence for methanogenic bacteria on an extrasolar planet would have profound implications for the origin of life in the universe.

ESA has similar plans for a free-flying mid-infrared space telescope intended to discover and characterize Earth-like planets: *Darwin*. The *Darwin* mission is currently planned for launch around 2015, prior to *TPF-I*. However, NASA and ESA have been discussing a joint *Darwin*/*TPF* mission for some time now, and considering the costs involved, a combined mission is an attractive possibility.

Once the *TPF*/*Darwin* missions fly and begin to send us images of faint dots of light orbiting around nearby solar-type stars, the push will be on to continue to the next logical step, and to consider building a new generation of space telescopes that will make the *TPF*/*Darwin* missions look like relatively modest adventures in comparison. Developing a space telescope that will be able to provide even crude images of the surface of an extrasolar Earth is a challenge that will not easily be met, but which will haunt us until we begin on the path toward its eventual development and deployment. Someday we will gaze at the oceans, clouds, and continents of other worlds, and wonder what it would be like to go there ourselves.

This work has been supported in part by the NASA Planetary Geology and Geophysics Program under grant NNG05GH30G and by the NASA Astrobiology Institute under grant NCC2-1056.

## REFERENCES

ALONSO, R., ET AL. 2004 *ApJ* **613**, L153.

BOND, I., ET AL. 2004 *ApJ* **606**, L155.

BOSS, A. P. 1998 *Looking for Earths: The Race to Find New Solar Systems.* J. Wiley & Sons.

BUTLER, R. P. & MARCY, G. W. 1996 *ApJ* **464**, L153.

BUTLER, R. P., ET AL. 2004 *ApJ* **617**, 580.

CHARBONNEAU, D., ET AL. 2000 *ApJ* **529**, L45.

CHARBONNEAU, D., ET AL. 2002 *ApJ* **568**, 377.

CHARBONNEAU, D., ET AL. 2005 *ApJ* **626**, 523.

CHAUVIN, G., ET AL. 2004 *A&A* **425**, L29.

DEMING, D., SEAGER, S., RICHARDSON, L. J., & HARRINGTON, J. 2005 *Nature* **434**, 740.

DUQUENNOY, A. & MAYOR, M. 1991 *A&A* **248**, 485.

GATEWOOD, G. & EICHHORN, H. 1973 *AJ* **78**, 769.

HENRY, G., ET AL. 2000 *ApJ* **529**, L-41.

KONACKI, M., ET AL. 2003 *Nature* **421**, 507.

MARCY, G. W. & BUTLER, R. P. 1996 *ApJ* **464**, L147.

MARCY, G. W., ET AL. 1997 *ApJ* **481**, 926.

MAYOR, M. & QUELOZ, D. 1995 *Nature* **378**, 355.

MCARTHUR, B. E., ET AL. 2004 *ApJ*, **614**, L81.

NEUHÄUSER, R., ET AL. 2005 *A&A* **435**, L13.

SANTOS, N. C., ET AL. 2004 *A&A*, **426**, L19.

SEAGER, S. & SASSELOV, D. D. 2000 *ApJ* **537**, 916.

SIGURDSSON, S., ET AL. 2003 *Science* **301**, 193.

SOZZETTI, A., ET AL. 2004 *ApJ* **616**, L167.

UDALSKI, A. ET AL. 2002 *Acta Astron.* **52**, 115.

VAN DE KAMP, P. 1963 *AJ* **68**, 515.

WALKER, G. A. H., ET AL. 1995 *Icarus* **116**, 359.

WOLSZCZAN, A. & FRAIL, D. A. 1992 *Nature* **355**, 145.

# The quest for very low-mass planets

By MICHEL MAYOR,† F. PEPE,
C. LOVIS, D. QUELOZ AND S. UDRY

Observatoire Astronomique de l'Université de Genève, 51 ch. des Maillettes,
1290 Sauverny, Switzerland

The Doppler technique has continuously improved its precision during the past two decades, attaining the level of $1\,\mathrm{m\,s^{-1}}$. The increasing precision opened the way to the discovery of the first extrasolar planet, and later, to the exploration of a large range of orbital parameters of extrasolar planets. This ability to detect and characterize in great detail companions down to Neptune-mass planets has provided many new and unique inputs for the understanding of planet formation and evolution. In addition, the success of the Doppler technique introduced a great dynamic in the whole domain, allowing the exploration of new possibilities.

Nowadays, the Doppler technique is no longer the only means to discover extrasolar planets. The performance of new instruments, like the High Accuracy Radial-velocity Planet Searcher (HARPS), has shown that the potential of the Doppler technique has not been exhausted; Earth-mass planets are now within reach. In the future, radial velocities will also play a fundamental role in the follow-up and characterization of planets discovered by means of other techniques—for transit candidates, in particular. We think, therefore, that the follow-up of candidates provided by, e.g., the *COnvection, ROtation and planetary Transits (COROT)* and *Kepler* space telescopes, will be of primary importance.

---

## 1. The success story of the Doppler technique

The advances made in the Doppler technique led to the discovery of brown dwarf companions of solar-like stars in the late '80s (Latham et al. 1989), reached (in 1995) into the hot-Jupiter domain with the discovery of the first planet orbiting a solar-type star (i.e., 51 Peg; Mayor & Queloz 1995), allowed the exploration of the full Saturn-to-Jupiter mass range up to several AU, and finally ended in 2004 with Neptune-mass candidates (Santos et al. 2004; McArthur et al. 2004; Butler et al. 2004b). As illustrated in Figure 1, the minimum masses of thus-discovered planetary companions decreased by almost three orders of magnitude in only 15 years. These huge advances go hand in hand with the continuous and remarkable increase of Doppler precision obtained on solar-like stars. This is best illustrated by the residuals of the fitted orbits to the data, as shown for some of the discovery milestones in Figure 2.

The impact of this amazing development has become evident during the past year: Table 1 shows planetary parameters of the seven Neptune-mass planets discovered in less than one year. Note the extremely low $m_2 \sin i$ of $5.9\,M_\oplus$ (Gl 876 d) and $m_2 \sin i / m_1$ of $4.2 \times 10^{-5}$ ($\mu$ Ara c) obtained on these objects. These discoveries not only confirm the existence of very low-mass planets around other solar-like stars, but also indicate that these objects must be very common. Therefore, new high-precision observations will help considerably to reduce the observational bias towards low masses and will deliver much data for the understanding of planet formation and evolution. Finally, we should remark that these recent results have made the astronomical community realize that Earth-like planets on short orbits are now within reach of the Doppler technique.

† Michel.Mayor@obs.unige.ch

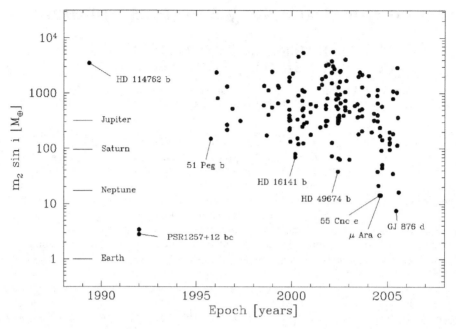

FIGURE 1. Evolution of the minimum mass of discovered exoplanets as a function of time.

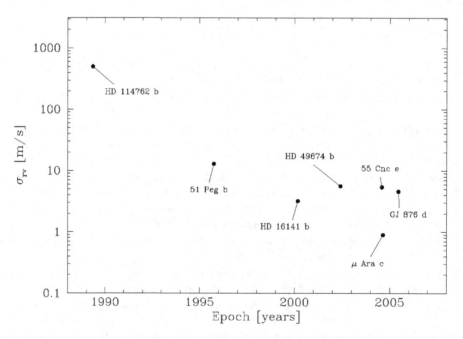

FIGURE 2. Evolution of the measurement residuals on orbits of discovered extrasolar planets as a function of time.

## 2. Is there any limit to the radial-velocity precision?

Only a couple of years ago, a part of the astronomical community believed that stellar noise would prevent reaching the $1\,\mathrm{m\,s^{-1}}$ precision level. Indeed, since such precision was not available then, the behavior of the stars below $3\,\mathrm{m\,s^{-1}}$ was completely unknown.

| Planet | P [days] | $m_2 \sin i$ [$M_\oplus$] | (o–c) [$m\,s^{-1}$] | $m_2 \sin i/m_1$ [$\times 10^{-5}$] | Reference |
|---|---|---|---|---|---|
| 55 Cnc e | 2.81 | 14 | 5.4 | 4.7 | McArthur et al. (2004) |
| $\mu$ Ara c | 9.6 | 14 | 0.9 | 4.2 | Santos et al. (2004) |
| Gl 436 b | 2.6 | 21 | 5.26 | 16.0 | Butler et al. (2004a) |
| HD 4308 b | 15.6 | 14 | 1.3 | 5.4 | Udry et al. (2005) |
| Gl 581 b | 4.96 | 16.6 | 2.5 | 17.1 | Bonfils et al. (2005) |
| Gl 876 d | 1.94 | 5.9 | 4.6 | 6.0 | Rivera et al. (2005) |
| HD 190360 c | 17.1 | 18 | 3.5 | 6.0 | Marcy et al. (2005) |

TABLE 1. Summary table of the recently discovered Neptune-mass planets. The lowest $m_2 \sin i$ of $5.9\,\mathrm{m\,s^{-1}}$ was obtained for Gl 876 d, while the lowest $m_2 \sin i/m_1$ of $4.2 \times 10^{-5}$ was achieved on $\mu$ Ara c.

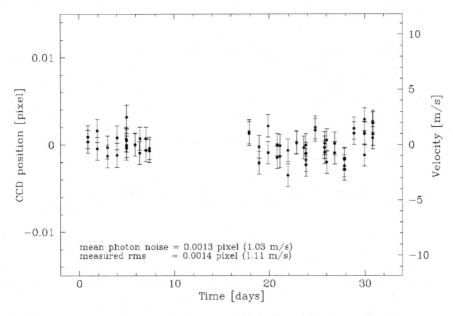

FIGURE 3. In order to illustrate the extraordinary stability of HARPS, we plot here the *measured physical* position of one single calibration spectrum line (thorium lamp) on the CCD as a function of time. The line position is expressed directly in the CCD pixel scale and does not include *any* calibration. It can be seen that the line remains at the same location on the CCD within $\sim 1\,\mathrm{m\,s^{-1}}$, which is the approximate photon-noise precision obtained on the centroid measurement of the spectral line. This diagram demonstrates that the instrument remained *intrinsically* stable during more than a month, without calibration!

However, with the increased precision of the HARPS instrument (Pepe et al. 2005), suddenly it became possible to explore this new radial-velocity range.

In the laboratory, we could easily prove that the instrumental stability of HARPS is unequaled (see Figure 3). During the instrument commissioning period, intense astero-seismological observations were carried out to monitor its short-term ($\sim$nights) precision. Besides demonstrating that the short-time precision was better than $20\,\mathrm{cm\,s^{-1}}$, these observations also revealed the fact that almost every star presents $p$-mode oscillations at this precision level. Such $p$-modes could clearly and directly be identified in the high-frequency measurement series obtained on several sample stars (Figure 4).

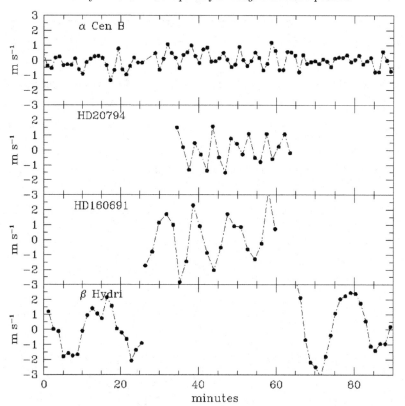

FIGURE 4. Summary plot of the radial-velocity measurements of stars for which we have made long observations series. All of them show clear indications of stellar pulsations.

From these measurements it immediately became evident that HARPS' capability for planet search was not limited by the instrument's performance, but rather by the stars themselves. Indeed, stellar $p$-mode oscillations on a short-term scale, and stellar jitter on a long-term scale, introduce obvious radial-velocity noise that cannot be mistaken at the precision level of HARPS. For instance, even a very 'quiet' G or K dwarf shows oscillation modes of several $10\,\mathrm{cm\,s^{-1}}$. These modes might combine and finally add up to radial-velocity amplitudes as large as several $\mathrm{m\,s^{-1}}$. Moreover, any exposure with an integration time shorter than the oscillation period of the star might arbitrarily fall on any phase of the pulsation cycle, leading to additional radial-velocity noise. This phenomenon may seriously compromise the ability to detect very low-mass planets around solar-type stars using radial-velocity measurements.

In June 2004, a group of European astronomers (Santos et al. 2004; Bouchy et al. 2005) proved that $p$-mode radial-velocity variability was not an issue for detecting very-low mass planets, providing some conditions are met. During asteroseismology campaigns on the star $\mu$ Ara, they measured its $p$-oscillation modes, but in addition, they also observed an unexpected coherent night-to-night variation in very small semi-amplitude of $4.1\,\mathrm{m\,s^{-1}}$. This was later confirmed to be the signature of a planetary companion of $m_2 \sin i = 14\,M_\oplus$ with an orbital period of $P = 9.6$ days (see Figure 5). This discovery demonstrated that oscillation noise can be averaged out to unveil a small radial-velocity signature of a Neptune-mass planet companion—if the total integration time chosen is sufficiently long when compared to the oscillation period. It must also be mentioned that

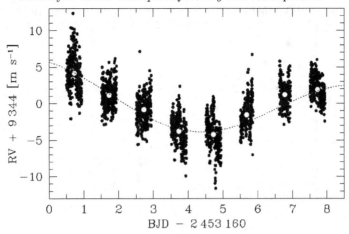

FIGURE 5. Asteroseismology observations of $\mu$ Ara. Although the dispersion of the radial velocity caused by stellar oscillations can rise $10\,\mathrm{m\,s^{-1}}$ amplitudes, one easily sees the 'low-frequency' variation induced by the planetary companion on the daily radial-velocity average.

the pulsations of $\mu$ Ara have high amplitudes up to $10\,\mathrm{m\,s^{-1}}$ $P-V$. This object represents, therefore, almost the worst case scenario if high precision is the aim. Despite this fact, the residuals could be decreased to the level of $0.4\,\mathrm{m\,s^{-1}}$ during the asteroseismology run (long integration time) and about $1\,\mathrm{m\,s^{-1}}$ during "ordinary" runs (see Figure 6).

The semi-amplitude of the radial-velocity wobble of $\mu$ Ara c-like objects is hardly larger than typical stellar $p$-mode oscillations. As mentioned above, the discovery of Neptune-mass planets may only be feasible when applying an adequate observation strategy that includes an integration time increased beyond the typical period of stellar oscillations (i.e., more than five minutes), in order to average them out. In practice, the total integration time was fixed to 15 minutes for all the stars of the high-precision HARPS sample, independent of the stellar magnitude. An example showing the success of this strategy is presented in Figure 7, which shows the low residuals of $0.9\,\mathrm{m\,s^{-1}}$ obtained on the radial-velocity curve of the planet-harboring star HD 102117 (Lovis et al. 2005).

We followed this strategy on a set of 200 selected stars of the HARPS planet-search program. For this very high-precision survey, an accuracy better than $1\,\mathrm{m\,s^{-1}}$ is desired for each individual measurement. So far, the results obtained demonstrate that this strategy is successful. The distribution histogram of the radial-velocity dispersion (Figure 8) peaks at $2\,\mathrm{m\,s^{-1}}$ and decreases rapidly towards higher values. More than 80% of these stars show dispersion smaller than $5.5\,\mathrm{m\,s^{-1}}$, and more than 35% have dispersions below $2.5\,\mathrm{m\,s^{-1}}$. It must be noted that the presented dispersion values include photon noise, stellar oscillations and jitter, and, in particular, radial-velocity wobble induced by known extrasolar planets ($\mu$ Ara c, HD 102117 b, HD 4308 b, etc.), or still-undetected planetary companions.

The increase in the performance of HARPS, combined with the new observation strategy, has led to discoveries of unequaled quality. This is best illustrated by the analyzing the residuals of the orbital fits to the measured data. Figure 9 shows the fit residuals for all extrasolar planets discovered since January 2004—about the time HARPS became fully operational. The planets discovered with HARPS are marked by the dashed area; those represented by the cross-dashed area are all part of the instrument's high-precision program. Note that *all* the planets with rms below $3\,\mathrm{m\,s^{-1}}$ have been discovered with

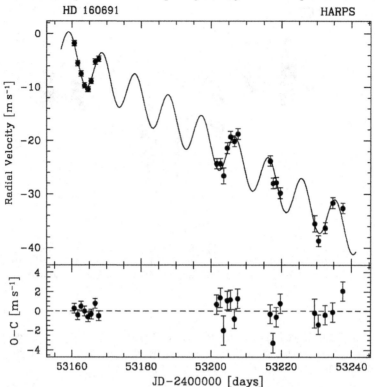

FIGURE 6. HARPS radial-velocity measurements of $\mu$ Ara as a function of time. The filled line represents the best fit to the data, obtained with the sum of a keplerian function and a linear trend, representing the effect of the long-period companions to the system. The residuals of the fit to the data, only 0.9 m s$^{-1}$ rms, are shown in the lower panel. Note that for the measurements obtained during the asteroseismology run (longer total integration), the residuals of the orbital fit to the data is as low as 0.4 m s$^{-1}$.

HARPS. This clearly demonstrates that the precision of HARPS is about a factor of two higher than any other existing instrument.

Since the discovery of the very low-mass companion of $\mu$ Ara, similar objects have been discovered (see Table 1). These planetary companions begin to populate the lower end of the secondary-mass distribution, a region so far affected by detection incompleteness. The discovery of these very low-mass planets close to the detection threshold of radial-velocity surveys suggests that this kind of object may be rather frequent. But just the simple existence of such planets could cause headaches for the theoretician. Indeed, statistical considerations predict that planets with a mass 1–0.1 $M_{\mathrm{Sat}}$ and with semi-major axis of 0.1–1 AU must be rare (Ida & Lin 2004, see also next section for details). At least for the moment, the recent discoveries contradict these predictions. In any case, the continuous detection of planets with even lower masses will set new constraints to possible planetary system formation and evolution models.

## 3. Constraining formation and evolution scenarios

Low-mass giant planets (i.e., planets with masses in the range 10–100 $M_{\oplus}$), are of particular interest as they provide potentially strong constraints on current giant planet formation and evolution models. Indeed, and perhaps contrary to intuition, the formation

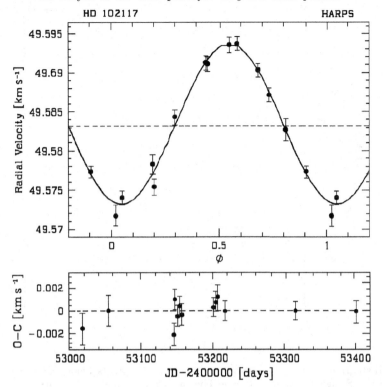

FIGURE 7. Measured radial velocity of HD 102117 in phase with the orbital period of the planet. The fitted orbital solution is shown as well. The residuals of the data points to this solution are only $0.9\,\mathrm{m\,s^{-1}}$ rms. This value includes photon noise and remaining 'stellar noise.'

of these objects within the current theoretical models appears more difficult than the formation of their more massive counterparts. For this reason, objects with masses within or at the edge of this range—like $\mu\,\mathrm{Ara\,c}$ and HD 4308 b (Udry et al. 2005)—are of particular interest.

In the direct-collapse scenario, planets form on very short timescales through gravitational collapse of patches of the proto-planetary disk (Boss 2002). High-resolution simulations of this process show that planets tend to form on elliptical orbits with semi-major axis of several astronomical units and masses between 1 and $7\,M_{\mathrm{Jup}}$ (Mayer et al. 2002, 2004). In this scenario, $\mu\,\mathrm{Ara}$-type planets would have to result from subsequent evolution involving migration and very significant mass loss.

In the framework of the core accretion model (e.g., Pollack et al. 1996), the final mass of a planet is actually determined by the amount of gas the core accretes after it has reached a critical mass, which is of order $10\text{–}15\,M_\oplus$. In Ida & Lin (2004), this amount is determined by the rate at which gas can be accreted (essentially the Kelvin-Helmholtz timescale) and by the total amount of gas available within the planet's gravitational reach. Since for super-critical cores (even in low-mass disks) the Kelvin-Helmoltz timescale is short and the amount of gas available large (compared to an Earth's mass), planets tend to form that are either less massive or significantly more massive than the critical mass. From a large number of formation model calculations, Ida & Lin (2004) found that only a very few planets form in the mass range $10\text{–}100\,M_\oplus$, a range they actually called a *planetary desert*.

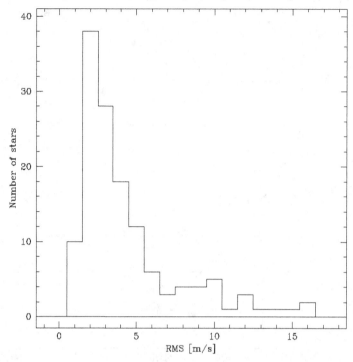

FIGURE 8. Histogram of the radial-velocity scatter for all targets belonging to the very high-
-precision HARPS planet-search program. The distribution peaks at $1.5\,\mathrm{m\,s^{-1}}$ and is mainly
dominated by stellar oscillations and jitter. We must point out here that $\mu\,\mathrm{Ara}$ (corrected for
the drift due to $\mu\,\mathrm{Ara\,b}$) and HD 102117, are part of this distribution, and that the orbital motions of their planets alone induce a radial-velocity scatter of about $2.5\,\mathrm{m\,s^{-1}}$ and $6\,\mathrm{m\,s^{-1}}$ rms,
respectively!

In the extended core accretion models of Alibert et al. (2005), due to the planet's
migration, it can in principle accrete gas over the entire lifetime of the disk. However,
since the latter thins out with time and the planet eventually opens a gap as it grows more
massive, the gas supply decreases with time. The growth rate of the planet is actually set
by the rate at which the disk can supply the gas, rather than the rate at which the planet
can accrete it. Monte-Carlo simulations are ongoing to verify whether these models lead
to a different planetary initial mass function as in Ida & Lin (2004).

At first glance, the relatively numerous small-mass objects discovered so far seem to
pose a problem to current planet-formation theories (Lovis et al. 2005)—however, the
situation is actually more complex. Since all the known very low-mass planets are located
close to their star, one cannot exclude the fact that these objects could have formed much
more massively and lost a significant amount of their mass through evaporation during
their lifetime (see Baraffe et al. 2004, 2005, for a more detailed discussion).

While mass loss from initially more massive objects could possibly account for the
light planets very close to their star, it is not clear whether $\mu\,\mathrm{Ara\,c}$—located at a distance
of 0.09 AU—could actually result from the evaporation of a more massive object. The
situation is even more critical for HD 4308 b (Udry et al. 2005), located further away
(0.115 AU) from its parent star which is, in addition, less luminous than $\mu\,\mathrm{Ara}$ by a
factor of $\sim1.8$. The effects could possibly be compensated, at least partially, by the very
old estimated age of the star. However, as more $\mu\,\mathrm{Ara}$-like objects are being discovered,
and if they all are the results of the evaporation of larger-mass planets, the question of the

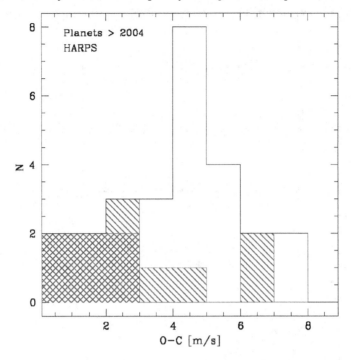

FIGURE 9. Histogram of the residuals of the best-orbit fit to the measured data for *all* extrasolar planets discovered since January 2004. The dashed planets have been discovered with HARPS, whereas the planet belonging to the HARPS Program Nr. 1 (high-precision) are distinguished by cross-dashed area.

probability of catching these systems shortly before complete evaporation will become a central one.

As already stated by Baraffe et al. (2005), the current evaporation models are still affected by large uncertainties—the lack of detailed chemistry treatment, non-standard chemical composition in the envelope, the effect or rocky/icy cores, etc.—that will need to be clarified in order to solve the question of the possible formation of $\mu$ Ara-type planets through evaporation.

Finally, given their close location to their star, the detected small-mass planets are likely to have migrated to their current position from further out in the nebula. The chemical composition of these planets will depend upon the extent of their migration, the thermal history of the nebula, and hence, the composition of the planetesimals along the accretion path of the planet. The situation is made more complicated by the fact that the ice-line itself is moving as the nebula evolves (see e.g., Sasselov & Lecar 2000). Detailed models of planetary formation including these effects have yet to be developed.

## 4. Detecting Earth-mass planets

The threshold of the lowest-mass planet detectable by the Doppler technique keeps decreasing. Today, with the currently achieved precision of about $1\,\mathrm{m\,s^{-1}}$, Neptune-mass planets can be discovered. Nobody has yet explored in detail the domain below the $1\,\mathrm{m\,s^{-1}}$ level. The measurements on $\mu$ Ara have demonstrated that it is possible to beat the stellar pulsation noise by investing sufficient observing time. However, one open issue remains unsolved: the behavior of the stars on longer timescales, where stellar jitter and

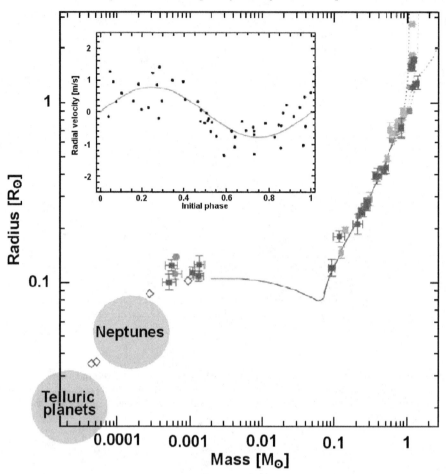

FIGURE 10. The mass-radius diagram from stars to planets. The low-mass end of the diagram can be investigated by the combination of precise transit search and radial-velocity measurements. With the combination of satellites like *COROT* and *Kepler* with a HARPS-like instrument, the Neptune- to Earth-mass domain will become accessible. The inserted diagram on the top left shows a simulation of the radial-velocity variation caused by two Earth-mass planets on a four-day orbit around a solar-type star. With the HARPS precision of about $60\,\mathrm{cm\,s^{-1}}$, only 50 measurements are needed to determine the planet mass with an accuracy within 10%.

spots may impact the final achievable accuracy. In this case, an accurate pre-selection of the stars may help focus on 'good' candidates and optimize the observation time. In addition, bisector analysis and follow-up of activity indicators such as $\log R_{\mathrm{HK}}$, as well as photometric measurements, would allow for the identification of potential error sources.

Nevertheless, to discover an extrasolar planet by means of the Doppler technique requires that the radial-velocity signal induced by the planet be significantly higher than the dispersion, or alternatively, requires that a large number of data points be recorded. This is particularly important to rule out artifacts, given the relatively high number of free parameters in the orbital solution, especially for multi-planet systems. A large number of measurements could overcome this problem, but would demand an enormous investment of observing time.

The many ongoing transit surveys may provide another interesting route to the characterization of very low-mass planets. If one considers a transit signal with a known orbital period, it is obvious that measuring its mass is less demanding, both on the number and the accuracy of the radial-velocity measurements. For example, a 2 $M_\oplus$ planet on a 4-day orbit would make a radial-velocity amplitude of about $80\,\mathrm{cm\,s^{-1}}$. Given the present precision of HARPS, which is estimated to about $60\,\mathrm{cm\,s^{-1}}$, it may be possible to detect and measure the amplitude of the radial-velocity wobble with only few radial-velocity measurements—provided that the period of the system is known in advance (see Figure 10 and included plot).

The *COROT* (2007–) and *Kepler* (2009–) space telescopes will provide many Neptune-sized and even Earth-sized planet candidates with orbits similar to, or smaller than, that of $\mu$ Ara c. The radial-velocity follow-up of these candidates will deliver their precise mass and orbit. When this information is combined with the transit-observation parameters, one obtains the mass-radius relation of planets in the domain of very low masses. This combined approach is currently being successfully carried out for the Optical Gravitational Lensing Experiment (OGLE) planetary candidates of about the mass and the size of Jupiter (Figure 10; see e.g., Pont et al. 2005). The almost mandatory radial-velocity follow-up of the *COROT* and *Kepler* planetary candidates is clearly within reach of the capabilities of a HARPS-like instrument. In this context, the most exciting aspect is the opportunity to explore mass-to-radius relation down to the Earth-mass domain.

## REFERENCES

ALIBERT, Y., MORDASINI, C., BENZ, W., & WINISDOERFFER, C. 2005 *A&A* **434**, 343.

BARAFFE, I., CHABRIER, G., BARMAN, T., ET AL. 2005 *A&A* **436**, L47.

BARAFFE, I., SELSIS, F., CHABRIER, G., ET AL. 2004 *A&A* **419**, L13.

BONFILS, X., ET AL. 2005 *A&A* **443**, L15.

BOSS, A. 2002 *ApJ* **576**, 462.

BOUCHY, F., BAZOT, M., SANTOS, N. C., VAUCLAIR, S., & SOSNOWSKA, D. 2005 *A&A* **440**, 609.

BUTLER, R. P., BEDDING, T. R., KJELDSEN, H., ET AL. 2004a *ApJ* **600**, L75.

BUTLER, R. P., VOGT, S. S., MARCY, G. W., ET AL. 2004b *ApJ* **617**, 580.

IDA, S. & LIN, D. 2004 *ApJ* **604**, 388.

LATHAM, D. W., STEFANIK, R. P., MAZEH, T., MAYOR, M., & BURKI, G. 1989 *Nature* **339**, 38.

LOVIS, C., MAYOR, M., BOUCHY, F., ET AL. 2005 *A&A* **437**, 1121.

MARCY, G. W., BUTLER, R. P., VOGT, S. S., ET AL. 2005 *ApJ*, **619**, 570.

MAYER, L., QUINN, T., WADSLEY, J., & STADEL, J. 2002 *Science* **298**, 1756.

MAYER, L., WADSLEY, J., QUINN, T., & STADEL, J. 2005 *MNRAS* **363**, 641.

MAYOR, M. & QUELOZ, D. 1995 *Nature* **378**, 355.

MCARTHUR, B. E., ENDL, M., COCHRAN, W. D., ET AL. 2004 *ApJ* **614**, L81.

PEPE, F., MAYOR, M., QUELOZ, D., ET AL. 2005 *The Messenger* **120**, 22.

POLLACK, J., HUBICKYJ, O., BODENHEIMER, P., ET AL. 1996 *Icarus* **124**, 62.

PONT, F., BOUCHY, F., MELO, C., ET AL. 2005 *A&A* **438**, 1123.

RIVERA, E., ET AL. 2005 *ApJ* **634**, 625.

SANTOS, N. C., BOUCHY, F., MAYOR, M., ET AL. 2004 *A&A* **426**, L19.

SASSELOV, D. & LECAR, M. 2000 *ApJ* **528**, 995.

UDRY, S., ET AL. 2006 *A&A* **447**, 361.

# Extrasolar planets: A galactic perspective

## By I. NEILL REID

Space Telescope Science Institute, 3700 San Martin Drive, Baltimore, MD 21218, USA

The host stars of extrasolar planets (ESPs) tend to be metal rich. We have examined other properties of these stars in search of systematic trends that might distinguish exoplanet hosts from the *hoi polloi* of the Galactic disk; we find no evidence for such trends among the present sample. The $\alpha$-element abundance ratios show that several ESP hosts are likely to be members of the thick disk population, indicating that planet formation has occurred throughout the full lifetime of the Galactic disk. We briefly consider the radial metallicity gradient and age-metallicity relation of the Galactic disk, and complete a back-of-the-envelope estimate of the likely number of solar-type stars with planetary companions with $6 < R < 10$ kpc.

## 1. Introduction

Exploitation of major scientific discoveries tends to follow a familiar pattern. Immediately following the initial discovery, the main focus is on verification—testing the initial analysis against alternative explanations for the same phenomena. The second phase centers on consolidation—acquiring additional observations and/or improved theoretical data on particular phenomena. The third phase is reached with the discovery of sufficient observational examples to map out a substantial fraction of the phase space occupied by key parameters.

After only a decade, investigations of extrasolar planets have clearly entered the third stage. More than 145 planetary-mass companions are known around more than 125 main sequence dwarfs and red giants. The overall properties of these planets are discussed elsewhere in this volume (see the review by Mayor). It is now well established that the frequency of (currently detectable) planetary systems increases with increasing metallicity of the parent star. Here, we concentrate on analyzing other salient properties of the extrasolar planetary (ESP) host stars, looking for evidence of additional potential correlations.

## 2. Wide planetary-mass companions of low-mass dwarfs

First, an aside: recent months have seen the imaging of low-mass companions to several young stars in the Solar Neighborhood. Neuhauser et al. (2005) have identified a faint, common proper motion (cpm) companion of GQ Lupi, a $\sim$0.45 $M_\odot$ member of the Lupus I group ($\tau \sim$ 1–2 Myrs); the companion lies at a separation of $\sim$100 AU and has a likely mass between 1 and 42 Jupiter masses ($M_J$). Similarly, Chauvin et al. (2005b) have discovered a faint cpm companion to AB Pic, a K2 dwarf, member of the $\sim$30 Myr-old Tucana-Horologium association; the companion is $\sim$245 AU distant, with a likely mass of 13–14 $M_J$. Finally, and most intriguingly, Chauvin et al. (2005a) have shown that 2MASS J1207334−393254, a $\sim$35 $M_J$ brown dwarf member of the $\sim$10-Myr-old TW Hydrae association, has a wide cpm companion, separation $\sim$60 AU, with a likely mass of only 2–5 $M_J$.

Should we classify these low-mass companions as planets or brown dwarfs? In my opinion, all of these objects are brown dwarfs. The mass estimates themselves, based on evolutionary models, offer no discrimination, since, while they overlap with the planetary régime, it remains unclear whether they lie beyond a lower limit to the brown dwarf mass

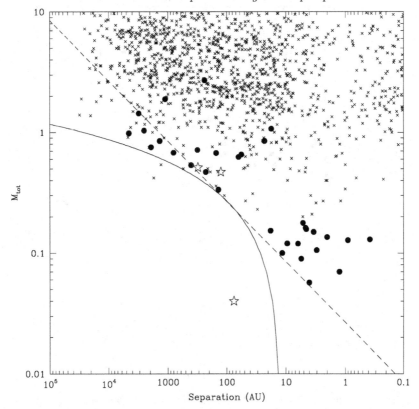

FIGURE 1. Wide planetary-mass companions of low-mass dwarfs: total mass as a function of separation in binary systems (adapted from Burgasser et al. 2003). The crosses plot data for stellar binaries, while solid points identify systems with brown dwarf components; the solid line outlines a linear relation between maximum separation and total mass, while the dotted line plots $a_{max} \propto M_{tot}^2$. The stars mark the location of 2M1207AB, GQ Lupi and AB Pic AB.

spectrum (if such exists). The large distance between each companion and its primary is a more telling parameter. There is little question that the Solar System planets formed from the Sun's protoplanetary disk; it seems reasonable to apply the same criterion in categorizing extrasolar planets. This test offers no discriminatory power at small separations, as with the 11 $M_J$ companion of HD 114762, which has an orbital semi-major axis of 0.3 AU. However, one might argue that disks are likely to have problems forming massive companions at large radii. Thus, it seems unlikely that a 35 $M_J$ brown dwarf would possess a protoplanetary disk sufficiently massive that it could form a 2–5 $M_J$ companion at a distance of 60 AU from the primary.

Nonetheless, this debate over terminology should not be allowed to obscure the significance of these discoveries, particularly 2M1207AB. Recent analyses show that the maximum separation of binaries in the field appears to correlate with the total mass (Burgasser et al. 2003). The newly discovered systems lie at the extremes of this distribution (Figure 1). 2M1207AB is particularly notable, with not only a projected separation, $\Delta$, four times higher than comparable brown dwarf binaries, but also a very low mass ratio, $q = \frac{M_2}{M_1} < 0.14$. All of these systems would be extremely difficult to detect at ages exceeding $\sim 10^8$ years; thus, their absence among field binaries could reflect either dynamical evolution (and system disruption), or observational selection effects.

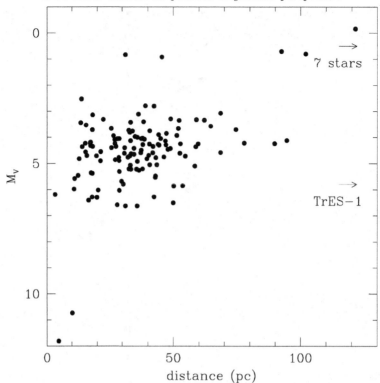

FIGURE 2. The distance distribution of stars known to have planetary-mass companions; absolute magnitudes are primarily based on Hipparcos parallax data.

## 3. The planetary host stars

The overwhelming majority of planetary systems have been discovered through radial velocity surveys, with a handful of recent additions from transit surveys and microlensing programs. Radial velocity surveys focus on solar-like stars, for obvious reasons, so it is not surprising that most of the ESP hosts are late-F to early-K dwarfs; moreover, almost all lie within 50 parsecs of the Sun (Figure 2).

Approximately 90% of the solar-type stars within 25 parsecs of the Sun are included in either the UC or Geneva surveys, although only the UC survey has published a catalog of non-detections (Nidever et al. 2001). Building on the local completeness of these surveys, Figure 3 shows the distribution of semi-major axes/projected separation as a function of mass for all known companions of solar-type stars within 25 parsecs of the Sun; the distribution clearly shows the brown dwarf desert, and provides the most effective demonstration that extrasolar planets are not simply a low-mass tail to the stellar/brown dwarf companion mass function.

### 3.1. *Stars and planetary properties*

Given the correlation between planetary frequency and metallicity, one might expect a bias towards higher mass planets in metal-rich stars; the current data, however, do not support that contention. This suggests that high metallicity acts as a trigger for planet formation, rather than playing a key role in the formation mechanism itself. Metal-rich systems ($[m/H] > 0.1$) do include a higher proportion of short-period, hot Jupiters, perhaps reflecting higher viscous drag in the protoplanetary disk (Sozzetti 2004; Boss, this conference).

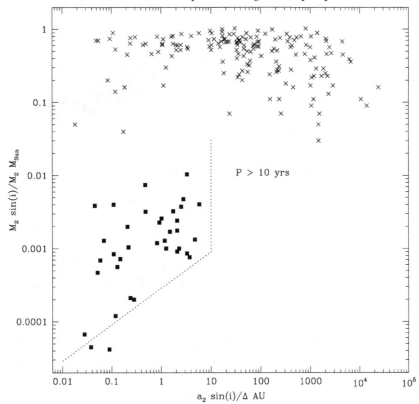

FIGURE 3. Companions mass as a function of separation for solar-type stars within 25 parsecs of the Sun. Crosses mark stellar and brown dwarf secondaries, while the solid squares identify planetary-mass companions. The dotted lines mark the effective limits of the planetary radial velocity surveys. The scarcity of systems with masses between 0.01 and 0.1 $M_\odot$ is the brown dwarf desert.

There is no obvious direct correlation between the mass of the primary star and the masses of planetary companions; for example, the $\sim$0.35 $M_\odot$ M3 dwarf, Gl 876, has two planets with masses comparable to Jupiter. On the other hand, one might expect an upper limit to the mass distribution to emerge, simply because lower mass stars are likely to have lower mass protoplanetary disks.

## 3.2. Metallicities and the thick disk

Chemical abundance, particularly individual elemental abundance ratios, serves as a population discriminant for the ESP host stars. Halo stars have long been known to possess $\alpha$-element abundances (Mg, Ti, O, Ca, Si) that are enhanced by a factor of 2–3 compared to the Sun. This is generally attributed to the short formation timescale of the halo (Matteucci & Greggio 1983): $\alpha$-elements are produced by rapid $\alpha$-capture, and originate in Type II supernovae, massive stars with evolutionary lifetimes of $10^7$ to $10^8$ years. In contrast, Type I supernovae, which are produced by thermal runaway on an accreting white dwarf in a binary system, have evolutionary timescales of 1–2 Gyrs. These systems produce a much higher proportion of Fe; thus, their ejecta drive down the $[\alpha/\text{Fe}]$ ratio in the ISM, and in newly forming stars.

Recent high-resolution spectroscopic analyses of nearby high velocity stars provide evidence that the thick disk is also $\alpha$-enhanced (Fuhrmann 1998, 2004; Prochaska 2000).

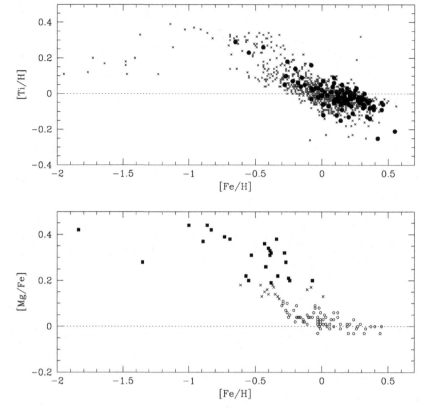

FIGURE 4. α-element abundances as a function of metallicity. The lower panel plots data for nearby stars from Fuhrmann (1998), where the open circles are disk dwarfs, the solid squares mark stars identified as members of the thick disk. The upper panel plots data from Valenti & Fischer's (2005) analysis of stars in the Berkeley/Carnegie planet survey; the solid points mark stars known to have planetary companions. Three stars (identified in the text) are almost certainly members of the thick disk.

The thick disk is the extended population originally identified from polar star counts by Gilmore & Reid (1983); current theories favor an origin through dynamical excitation by a major merger early in the history of the Milky Way (Bensby 2004). With abundances in the range $-1 < [m/H] < -0.3$ (Figure 4, lower panel), the thick disk clearly formed after the Population II halo, but before Type I supernovae were able to drive up the iron abundance. Thus, the thick disk population almost certainly comprises stars from the original Galactic disk, which formed within the first 1–2 Gyrs of the Milky Way's history.

What is the relevance of these observations to planet formation? Valenti & Fischer (2005) have recently completed abundance analysis of the high-resolution spectroscopic data acquired by the Berkeley/Carnegie radial-velocity survey. Their analysis includes measurement of the abundance of Ti, an α-element. The upper panel of Figure 4 shows the distribution of the full sample as a function of [Fe/H], identifying stars known to have planets. Three of the latter stars have α-abundances consistent with thick disk stars (HD 6434, 0.48 $M_J$ planet; HD 37124, 0.75 $M_J$; HD 114762, 11 $M_J$), while three other stars (HD 114729, 0.82 $M_J$; ρ CrB, 1.04 $M_J$; HD 168746, 0.23 $M_J$) have intermediate values of [α/Fe]. These results strongly suggest that, even though planets may be extremely rare among metal-poor halo stars (Gilliland et al. 2000), planetary systems have been forming in the Galactic disk since its initial formation.

Geneva 40 pc vs. planet hosts

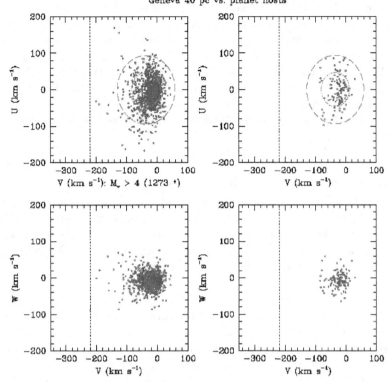

FIGURE 5. The velocity distribution of ESP host stars (right hand panels) compared with the velocity distribution of solar-type stars within 40 parsecs of the Sun.

### 3.3. *Kinematics*

Stellar kinematics are usually characterized using the Schwarzschild velocity ellipsoid; probability plots (Lutz & Upgren 1980) allow one to compare stellar samples that include multiple components with Gaussian velocity distributions (see, for example, Reid, Gizis & Hawley 2002). With the completion of the Geneva-Copenhagen survey of the Solar Neighborhood (Nordström et al. 2004), we have distances, proper motions, radial velocities and abundances† for most solar-type stars within 40 parsecs of the Sun. There are 1,273 stars within that distance that have $0 < (b-y) < 0.54$ and $M_V \geqslant 4.0$; analyzing their kinematics, we have

$$(U,\ V,\ W; \sigma_U,\ \sigma_V,\ \sigma_W) = (-9.6, -20.2, -7.6;\ 38.8,\ 31.0,\ 17.0\ \mathrm{km\ s}^{-1})\ , \qquad (3.1)$$

with an overall velocity dispersion, $\sigma_{\mathrm{tot}} = 52.7\ \mathrm{km\ s}^{-1}$.

There are 129 ESP hosts with accurate distances and space motions, and their mean kinematics are

$$(U,\ V,\ W;\ \sigma_U,\ \sigma_V,\ \sigma_W) = (-4.0, -25.5, -20.4;\ 37.7,\ 22.9,\ 20.4\ \mathrm{km\ s}^{-1})\ , \qquad (3.2)$$

with an overall velocity dispersion, $\sigma_{\mathrm{tot}} = 48.6\ \mathrm{km\ s}^{-1}$. All three stars identified as likely members of the thick disk (and two of the possible members) have high velocities (60 to 110 km s$^{-1}$) relative to the Sun.

---

† One should note that the abundances in this catalog are tied to the Schüster et al. *uvby*-based metallicity scale, which has color-dependent systematic errors (Haywood 2002). Those systematic errors can be corrected.

The two velocity distributions are very similar. The total velocity dispersion of the ESP hosts is slightly lower than the field, mainly reflecting the higher proportion of old, metal-poor stars in the latter population (this also accounts for the lower value of $\sigma_V$ for the ESP hosts). Interpreted in terms of the standard stellar diffusion model, $\sigma_{\text{tot}} \propto \tau^{1/3}$, the observed difference formally corresponds to a difference of only 20% in the average age. The relatively high mean motion perpendicular to the Plane is somewhat surprising, and may warrant further investigation. With that possible exception, however, the velocity distribution of ESP host stars is not particularly unusual, given the underlying metallicity distribution.

## 4. Planetary cartography in the Galaxy

The ultimate goal of these statistical analyses is to answer a simple question: How many stars in the Galaxy are likely to have associated planetary systems? While we are still far from being able to construct a reliable detailed model for the stellar populations in the Milky Way, there have been some initial studies, notably by Gonzalez, Brownlee, & Ward (2001). Here, we briefly consider those models.

### 4.1. *Abundance gradients and the Galactic Habitable Zone*

Planetary habitability is likely to depend on many factors: distance from the parent star; the stellar luminosity, activity and lifetime; (perhaps) planetary mass, which may depend on the metallicity of the system; (perhaps) the presence of a massive satellite; the existence of plate tectonics; and (perhaps) the space motion of the system. In contrast, from the perspective of planet formation, the correlation between planetary frequency and metallicity is the only significant factor that has emerged from statistical analysis of the known ESP systems. Extending the present results beyond the Solar Neighborhood requires modeling of both the radial abundance gradient and the age-metallicity relation of the Galactic disk.

Most studies assume a logarithmic radial abundance gradient; for example, Gonzalez et al. (2001) adopt a gradient,

$$\frac{\delta[m/H]}{\delta R} ,$$

of $-0.07$ dex kpc$^{-1}$. More recent analyses of intermediate-age Cepheids (Andrievsky et al. 2002) suggest a more complex radial distribution, with a flatter gradient in the vicinity of the Solar Radius, but steeper gradients outwith these limits. Figure 5a matches both distributions against [O/H] abundances for H II regions (from Shaver et al. 1983); we also show current estimates for the Sun, and for the Hyades, Pleiades and Praesepe clusters. By and large, the data favor the complex gradient, but one should note that extrapolating the inner gradient implies unreasonably high metallicities at $R < 4$ kpc. It is probably more reasonable to infer that we require more observations of metallicity tracers in the inner Galaxy.

The metallicity gradients plotted in the upper panel of Figure 6 are based on young objects—even the Cepheids are less the $10^9$ years old. Thus, the distributions are characteristic of the average metallicity of stars forming in the present-day Galaxy. Clearly, there is a substantial distribution of metallicity among the stars in the Solar Neighborhood, and one would expect comparable dispersions at other radii. Gonzalez et al. (2001) address this issue by assuming an age-metallicity relation,

$$\frac{\delta[m/H]}{\delta\tau} = 0.035 \text{ dex Gyr}^{-1} ;$$

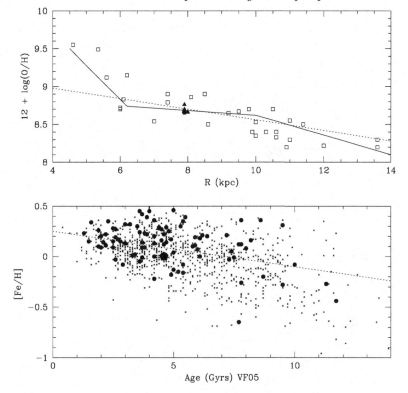

FIGURE 6. The upper panel plots oxygen abundance data for H II regions as a function of Galactic radius (open squares, data from Shaver et al. 1983). We also show the oxygen abundances for the Sun (solid square) and for the three nearest open clusters, the Hyades, Pleiades and Praesepe (solid triangles). The dotted line plots the abundance gradient adopted by Gonzalez et al. (2001); the solid line marks the composite gradient derived by Andrievsky et al. (2002) from Cepheid data. The lower panel plots the age-metallicity distribution derived by Valenti & Fischer (2005) for stars in the Berkeley/Carnegie radial velocity survey; solid points mark stars known to be ESP hosts. The pentagon marks the Sun.

they assume a small dispersion in metallicity (<0.08 dex). Note that this type of relation corresponds to a constant fractional increase in metallicity with time; that, in turn, requires either an increasing yield with increasing metallicity or an increasing star formation rate at a constant yield.

The lower panel of Figure 6 matches the Gonzalez et al. age-metallicity relation against empirical results from Valenti & Fischer's (2005) analysis of stars in the Berkeley/Carnegie survey; both ESP hosts and the Sun are separately identified. Unlike the classic Edvardsson et al. (1993) analysis, there is a trend in mean abundance with age. Overall, the observations indicate a much broader dispersion in metallicity, at all ages, than assumed by Gonzalez et al. (2001).† As with the radial abundance gradient, further analysis is required before settling on reliable values of these parameters.

### 4.2. *Planetary statistics in the Solar Circle*

Given all the caveats, what can we say about the likely distribution of planetary systems in the Milky Way? For the moment, any calculations must be restricted to regions rela-

---

† Note that the Sun, which lies at the mode of the local abundance distribution, appears somewhat metal poor for its age, as compared with VF05 data.

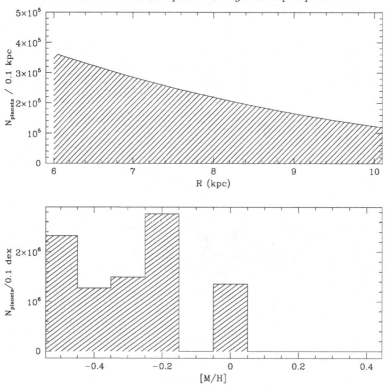

FIGURE 7. The upper panel plots the expected numbers of ESP hosts as a function of Galacto-centric distance for $6 < R < 10$ kpc; the lower panel plots the corresponding predicted metallicity distribution.

tively close to the Sun. However, Figure 6 suggests that the local metallicity distribution may well be representative of stars with Galactic radii between 6 and 10 kiloparsecs. Taking this as a working assumption, we can calculate a back-of-the-envelope estimate of how many solar-type stars in this annulus might have planetary companions.

In making this calculation, we identify solar-type stars as stars with luminosities $4 < M_v < 6$; we assumed a density law

$$\rho(R) = \rho_0 e^{(R-R_0)/h} e^{-z/z_0} \quad , \tag{4.1}$$

where $\rho_0 = 4.4 \times 10^{-3}$ stars pc$^{-3}$, with 90% of the stars assigned to the disk and 10% to the thick disk; $h = 2500$ pc; and $z_0 = 300$ pc for the disk, and $z_0 = 1000$ pc for the thick disk. We assume an overall planetary frequency of 6%.

Given those starting assumptions, Figure 7 shows the expected radial density distribution and the metallicity distribution of ESP hosts. In total, we predict that $\sim 3.5 \times 10^7$ solar-type stars are likely to have gas giant planetary companions with $a < 4$ AU. The space density increases with decreasing Galactocentric radius (density wins over surface area), and the metallicity distribution is approximately flat for $-0.1 < [m/H] < 0.3$. ESP hosts are not uncommon in the Solar Circle.

## 5. Summary and conclusions

More than 130 stars are now known to harbor planetary systems. The correlation between planetary frequency and metallicity is now well established, and probably reflects nature rather than nurture. We have analyzed the statistical properties of the current sample of ESP hosts, looking for other possible trends and biases. With the possible exception of a higher mean velocity perpendicular to the Plane, the planetary hosts appear to be unremarkable members of the Galactic Disk.

Several ESP host stars show enhanced $\alpha$-element abundances and high velocities relative to the Sun, strongly suggesting that they are members of the thick disk. If so, this indicates that planet formation has been a constant presence in the Galactic Disk.

We briefly considered the radial abundance gradient and age-metallicity distribution of disk stars. Recent results suggest that the gradient is flatter in the vicinity of the Sun than the canonical value, but probably steepens at larger radii; there is relatively little reliable data for the inner disk, $R < 6$ kpc. The age-metallicity distribution of the Berkeley/ Carnegie RV sample shows substantial dispersion at all ages. While the uncertainties in these parameters limit our ability to model the full Galaxy, we can use local statistics to estimate the planetary population in the vicinity of the Sun. We estimate that there are over $3.5 \times 10^7$ 'RV-detectable' planetary systems with Galactocentric radii in the range 6 to 10 kpc.

Thanks to Jeff Valenti for early access to computer-readable tables listing the results of the spectroscopic analysis of the Berkeley/Carnegie radial velocity sample.

## REFERENCES

ANDRIEVSKY, S. M., KOVTYUKH, V. V., LUCK, R. E., LÉPINE, J. R. D., MACIEL, W. J., & BELETSKY, Y. V. 2002 *A&A* **392**, 491.

BENSBY, T. 2004 *Ph.D. Thesis*.

BURGASSER, A. J., KIRKPATRICK, J. D., REID, I. N., BROWN, M. E., MISKEY, C. L., GIZIS, J. E. 2003 *ApJ* **586**, 512.

CHAUVIN, G., LAGRANGE, A.-M., DURNAS, C., ZUCKERMAN, B., MOUILLET, D., SONG, I., BEUZIT, J.-L., LOWRANCE, P. 2005a *A&A* **438**, L25.

CHAUVIN, G., LAGRANGE, A.-M., ZUCKERMAN, B., DURNAS, C., MOUILLET, D., SONG, I., BEUZIT, J.-L., LOWRANCE, P., BESSELL, M. S. 2005b *A&A* **438**, L29.

EDVARDSSON, B., ANDERSEN, J., GUSTAFSSON, B., LAMBERT, D. L., NISSEN, P. O., TOMKIN, J. 1993 *A&A* **275**, 101.

FUHRMANN, K. 1998 *A&A* **338**, 161.

FUHRMANN, K. 2004 *Astron. Nact.* **325**, 3.

GILLILAND, R. L., ET AL. 2000 *ApJ* **545**, L47.

GILMORE, G. F. & REID, I. N. 1983 *MNRAS* **202**, 1025.

GONZALEZ, G., BROWNLEE, D., & WARD, P. 2001 *Icarus* **152**, 185.

HAYWOOD, M. 2002 *MNRAS* **337**, 151.

LINEWEAVER, C. H., FENNER, Y., & GIBSON, B. K. 2004 *Science* **303**, 59.

LUTZ, T. E. & UPGREN, A. R. 1980 *AJ* **85**, 1390.

MATTEUCCI, F. & GREGGIO, L. 1986 *A&A* **154**, 279.

NEUHÄUSER, R., GUENTHER, E. W., WUCHTER, G., MUGRAUER, M., BEDALOV, A., & HAUSCHILDT, P. H. 2005 *A&A* **435**, 113.

NIDEVER, D. L., MARCY, G. W., BUTLER, R. P., FISCHER, D. A., & VOGT, S. S. 2002 *ApJS* **141**, 503.

NORDSTRÖM, B., ET AL. 2004 *A&A* **418**, 989.

PROCHASKA, J. X., NAUMOV, S. O., CARNEY, B. W., MCWILLIAM, A., & WOLFE, A. M. 2000 *AJ* **120**, 2513.

REID, I. N., GIZIS, J. E., & HAWLEY, S. L. 2002 *AJ* **124**, 2721.

SHAVER, P. A., MCGEE, R. X., NEWTON, L. M., DANKS, A. C., & POTTASCH, S. R. 1983 *MNRAS* **204**, 53.

SOZZETTI, A. 2004 *MNRAS* **354**, 1194.

VALENTI, J. A. & FISCHER, D. A. 2005 *ApJS* **159**, 141.

# The *Kepler* mission: Design, expected science results, opportunities to participate

By WILLIAM J. BORUCKI,[1] DAVID KOCH,[1]
GIBOR BASRI,[2] TIMOTHY BROWN,[3]
DOUGLAS CALDWELL,[4] EDNA DEVORE,[4]
EDWARD DUNHAM,[5] THOMAS GAUTIER,[6]
JOHN GEARY,[7] RONALD GILLILAND,[8]
ALAN GOULD,[9] STEVE HOWELL,[10] JON JENKINS,[4]
AND DAVID LATHAM [7]

[1]NASA Ames Research Center, Moffett Field, CA 94035, USA;
William.J.Borucki@nasa.gov; David.G.Koch@nasa.gov

[2]University of California, Berkeley, CA 94720, USA; Gibor@astron.berkeley.edu

[3]High Altitude Observatory, NCAR, Boulder, CO 80307, USA; TimBrown@hao.ucar.edu

[4]SETI Institute, Mountain View, CA 94043, USA;
dcaldwell@mail.arc.nasa.gov; EdeVore@seti.org; jjenkins@mail.arc.nasa.gov

[5]Lowell Observatory, Flagstaff, AZ 86001, USA; Dunham@lowell.edu

[6]Jet Propulsion Laboratory, Pasadena, CA 91109, USA; Thomas.N.Gautier@jpl.nasa.gov

[7]Harvard Smithsonian Center for Astrophysics, Harvard, MA 02138, USA;
geary@cfa.harvard.edu; dlatham@cfa.harvard.edu

[8]Space Telescope Science Institute, Baltimore, MD 21218, USA; gillil@stsci.edu

[9]Lawrence Hall of Science, University of California, Berkeley, CA 94720, USA;
AGould@uclink.berkeley.edu

[10]University of California, Riverside, CA 92521, USA; Steve.Howell@ucr.edu

*Kepler* is a Discovery-class mission designed to determine the frequency of Earth-size and smaller planets in and near the habitable zone (HZ) of spectral type F through M dwarf stars. The instrument consists of a 0.95 m aperture photometer to do high-precision photometry of 100,000 solar-like stars to search for patterns of transits. The depth and repetition time of transits provide the size of the planet relative to the star and its orbital period. Multi-band ground-based observation of these stars is currently underway to estimate the stellar parameters and to choose appropriate targets. With these parameters, the true planet radius and orbit scale—hence the relation to the HZ—can be determined. These spectra are also used to discover the relationships between the characteristics of planets and the stars they orbit. In particular, the association of planet size and occurrence frequency with stellar mass and metallicity will be investigated. At the end of the four-year mission, several hundred terrestrial planets should be discovered with periods between 1–400 days, if such planets are common. A null result would imply that terrestrial planets are rare. Based on the results of the recent Doppler-velocity discoveries, over a thousand giant planets will also be found. Information on the albedos and densities of those giants showing transits will be obtained. The mission is now in Phase C/D development and is scheduled for launch in 2008 into a 372-day heliocentric orbit.

## 1. Introduction

Since the first discoveries of planetary companions around normal stars in 1995, more than 150 such planets have been discovered. At least 5%, and as many as 25%, of solar-like stars show the presence of giant planets (Lineweaver & Grether 2003). These planets are generally very massive, often exceeding that of Jupiter and Saturn. Further, most have semi-major axes less than 1 AU and have high orbital eccentricities. The surprisingly small values for the semi-major axes imply that they form at several AU, but then lose momentum to the accretion disk and spiral inward. It is unclear what processes terminate the inward motion and what fraction of planets fall into the star. However, it is obvious that the inward motion of the giant planets will remove smaller planets by scattering them either into the star or out of the planetary system. It is also possible that the stars not showing the presence of giant planets are devoid of all planets because the giant planets merged with the star after their inward migration. Thus, planetary systems with terrestrial planets might be very rare. However, a recent discovery (Rivera et al. 2005) of a planet with a mass 7.5 times that of the Earth—with a 2.7-day orbital period—indicates that at least some terrestrial planets survive.

Determination of the frequency of terrestrial planets (and their distributions of size and orbital semi-major axes) is needed to increase our understanding of the structure of planetary systems. The *Kepler* mission is designed to discover hundreds of terrestrial planets in and near the habitable zone (HZ) around a wide variety of stars. For short-period orbits, hundreds of transits will be observed during the four-year mission so that planets as small as Mercury and Mars can be detected. *Kepler* is a PI-lead mission and was competitively selected as NASA Discovery Mission #10 in December 2001. It is scheduled to launch in June 2008 into an Earth-trailing orbit. A description of the mission and the expected science results are presented.

## 2. Scientific goals

The general scientific goal of the *Kepler* mission is to explore the structure and diversity of planetary systems, with special emphasis on determining the frequency of Earth-size planets in the HZ of solar-like stars. This is achieved by surveying a large sample of stars to:

• Determine the frequency of terrestrial-size ($R_\oplus$) and larger planets in or near the habitable zone of a wide variety of spectral types of stars;

• Determine the distributions of sizes and orbital semi-major axes of these planets;

• Estimate the frequency of planets orbiting multiple-star systems;

• Determine the distributions of semi-major axis, eccentricity, albedo, size, mass, and density of short period giant planets;

• Identify additional members of each photometrically-discovered planetary system using complementary techniques; and

• Determine the properties of those stars that harbor planetary systems.

The *Kepler* mission supports follow-on missions such as SIM, *TPF*, and the *Darwin* mission by finding the association between the frequency and characteristics of terrestrial planets and stellar type and by determining the distributions of planetary size and orbital semi-major axis. Since the *Kepler* field-of-view (FOV) is along a galactic arm at the same galactocentric distance as the Sun, the stellar population sampled with *Kepler* is indistinguishable from the immediate solar neighborhood. Thus, the *Kepler* results should provide pertinent information for selecting the most promising targets for future missions.

## 3. Photometer and spacecraft description

The instrument is a wide-FOV differential photometer with a 100 square degree FOV that continuously and simultaneously monitors the brightness of 100,000 main-sequence stars with sufficient precision to detect transits by Earth-size planets orbiting G2 dwarfs.

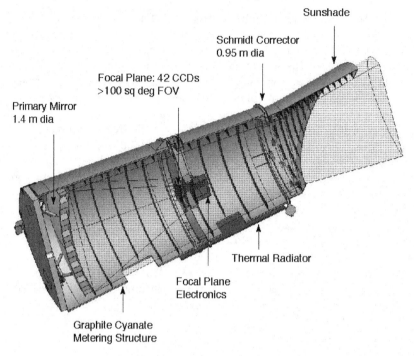

FIGURE 1. Isometric view of the *Kepler* photometer.

The brightness range of target stars is from visual magnitude 9 through 15. The photometer is based on a modified Schmidt telescope design that includes field-flattener lenses near the focal plane. Figure 1 is an isometric view of the photometer. The corrector has an aperture of 0.95 m with a 1.4 m diameter F/1 primary. This aperture is sufficient to reduce the Poisson noise to the level required to obtain a $4\sigma$ detection for a single transit from an Earth-size planet transiting a 12th magnitude G2 dwarf with a 6.5 hour transit. The focal plane is composed of 42 $1024 \times 2200$ backside-illuminated CCDs with 27 $\mu$m pixels.

The detector focal plane is at prime focus and is cooled by heat pipes that carry the heat out to a radiator in the shadow of the spacecraft. The low-level electronics are placed immediately behind the focal plane. A four-vane spider supports the focal plane and its electronics and contains the power- and signal-cables and the heat pipes.

The spacecraft bus encloses the base of the photometer and supports the arrays and the communication, navigation, and power equipment. Two antennas with different frequency coverage and gain patterns are available for uplink commanding and for data downlink. A steerable high-gain antenna operating at $Ka$ band is used for high-speed data transfer to the Deep Space Network (DSN). It is the only articulated component other than the ejectable cover. Approximately one GByte/day of data are recorded and then transferred to the ground every few days when contact is made with the DSN. The spacecraft provides very stable pointing using four fine guidance sensors mounted in the photometer focal plane. Small thrusters are used to desaturate the momentum wheels. Sufficient expendables are carried to extend the mission to six years.

Both the photometer and the spacecraft are being built by the Ball Aerospace and Technology Corporation (BATC) in Boulder, Colorado. NASA Ames manages the photometer development, mission and operations, and scientific analysis. JPL manages space-

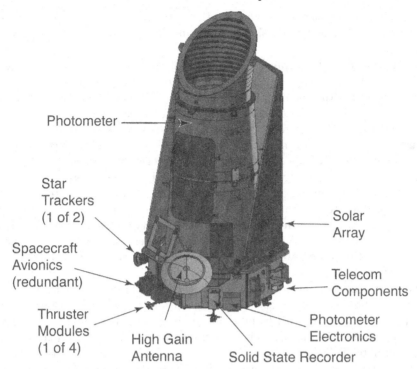

Photometer

Star
Trackers
(1 of 2)

Solar
Array

Spacecraft
Avionics
(redundant)

Telecom
Components

Thruster
Modules
(1 of 4)

High Gain
Antenna

Photometer
Electronics

Solid State Recorder

FIGURE 2. Integrated spacecraft and photometer.

craft and mission development. A more comprehensive discussion of the mission design is given in Koch et al., found in the poster book for this symposium.

## 4. Scientific approach

To achieve the required photometric precision to find terrestrial-size planets, the photometer and the data analysis system must be designed to detect the very small changes ($\sim$1 part in $10^4$) in stellar flux that are characteristic of transits by terrestrial planets. The *Kepler* mission approach is best described as "differential relative photometry." In this approach:

- Target stars are always measured relative to the ensemble of similar stars on the same part of the same CCD and read out by the same amplifier;
- Only the time change of the ratio of the target star to the ensemble is of interest. Only decreases from a trend line based on a few times the transit duration are relevant (long-term stability of the trend is not required);
- Target star and ensemble stars are read out every five seconds to avoid drift and saturation; and
- Correction for systematic errors is critical.

Photometry is not done on the spacecraft. Instead, all of the pixels associated with each star image and the collateral, bias, and smear pixels are sent to the ground for analysis. This choice allows many different approaches to be used to reduce systematic errors.

The spacecraft will be placed in an Earth-trailing heliocentric orbit by a Delta II 2925-10L launch vehicle. The heliocentric orbit provides a benign thermal environment to maintain photometric precision. It also allows continuous viewing of a single FOV for the entire mission without the Sun, Earth or Moon obtruding. Only a single FOV is monitored during the entire mission to avoid missing transits and to maintain a high duty cycle.

A pattern of at least three transits, showing that the orbital period repeats to a precision of at least 10 ppm and showing at least a $7\sigma$ detection, is required to validate any discovery. A detection threshold of $7\sigma$ is required to avoid false positives due to random noise. To obtain a higher recognition rate, the mission is designed to provide a lifetime of four years to allow four transits in the HZ of a solar-like star to be observed. Note that transit signatures with a mean detection statistic of $8\sigma$ will be recognized 84% of the time, whereas those with a mean of $7\sigma$ will be recognized only 50% of the time.

Classical signal-detection algorithms that whiten the stellar noise, fold the data to superimpose multiple transits, and apply matched filters are employed to search for the transit patterns down to the statistical noise limit (6). From measurements of the period, change in brightness and known stellar type, the planetary size, and the semi-major axis, the characteristic temperature of the planet can be determined. The latter gives some indication of whether liquid water could be present on the surface; i.e., whether the planet is in the habitable zone.

Because of the limitations on the telemetry stream, only data from those pixels illuminated by the pre-selected target stars are saved for transmission to Earth. Data for each pixel are co-added onboard to produce one brightness measurement per pixel per 15-minute integration. Data for a subset of target stars can be measured at a cadence of once per minute. This option will be exercised to obtain detailed emersion-immersion profiles, for detecting changes in transit timing due to the presence of multiple planets, and for conducting observations for astroseismology.

## 5. Selection of target stars and field of view

Continuously monitoring approximately 100,000 quiet, late-type target stars will provide a statistically meaningful estimate of the frequency of terrestrial planets in the HZ of solar-like stars. Centered on a galactic longitude of 70° and latitude of +6°, the FOV satisfies both the constraint of a 55° sun-avoidance angle and provides a very rich star field. This FOV falls within the Cygnus-Lyra constellations and results in looking in a tangential direction from the galactic center. In the 100 square degree *Kepler* FOV, there are approximately 450,000 stars brighter than 15th magnitude. A ground-based observation program led by David Latham (SAO) and Tim Brown (HAO) is underway to observe $2 \times 10^6$ stars in the FOV brighter than 17th magnitude. A unique color-filter system, based on the Sloan system and augmented with special filters, is used to identify both the luminosity class and spectral type of each star. Ancillary information from the 2 Mass catalog is also used. The resulting catalog allows the *Kepler* mission to choose only F through M dwarfs and to exclude giants and early spectral types from the target list. By classifying the stars for which we have complete photometry and all three 2 Mass bands available down to $K = 14.5$th magnitude, several thousand M dwarfs can be found and put on the target list. Because of their small diameter, these stars will provide sufficient signal-to-noise ratio (SNR) for detection of terrestrial-size planets even though they are dimmer than the majority of other targets. The *Kepler* results for the frequency of terrestrial planets orbiting M dwarfs are important because most nearby stars are M dwarfs.

Stellar variability sets the limit to the minimum size of planet that can be detected. It reduces the signal detectability in two important ways:

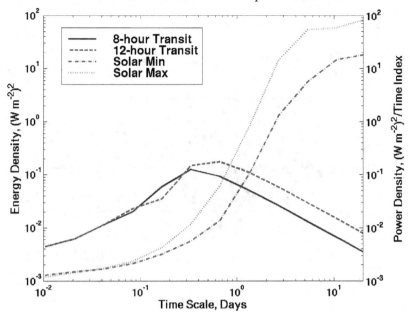

FIGURE 3. Power spectra of solar variability at solar maximum and minimum. Also shown are energy spectra of 8-hr and 10-hr transits (from Jenkins 2002.)

• The variability introduces noise into the detection passband and thereby reduces the SNR, and thus the statistical significance, of transits.

• Because the flux of every target star is ratioed to the fluxes of several surrounding stars to reject common-mode instrument noise, variability of the stars used in the normalization introduces noise into the target-star signal.

The second concern can be alleviated by measuring the variability of each star relative to an ensemble of others and then iteratively removing the noisiest from the list of comparison standards. To mitigate the effects of the first concern, stars must be chosen that have low variability.

Power spectra for the Sun at solar maximum and minimum are shown in Figure 3.

Also shown are the energy spectra for transits with 8- and 10-hour durations. It's clear that most of the solar variability is at periods substantially longer than those associated with planetary transits. In particular, the Sun's variability for samples with duration similar to that for transits is about 10 ppm. For stars rotating more rapidly than the Sun, the power spectrum will increase in amplitude and move to shorter periods, thus increasing the noise in the detection passband.

Stellar variability in late-type main-sequence stars is usually associated with the interplay of the convective layer and the internal magnetic field. Because the depth of the convective layer is a function of the spectral type of the star, and because the activity level is higher when the star is rotating rapidly, the variability of solar-like main-sequence stars is related to both their spectral type and rotation rate. Further, because the rotation rate decreases with age, the age of a star is an important variable. Thus, we expect that the factors that influence the variability of target stars are age and spectral type.

The age and rotation rate of the Sun are approximately 5 Gyr and 27 days, respectively. The age of the Galaxy is about 13.7 Gyr; about two-thirds of the stars are older than the Sun and are expected to be at least as quiet as the Sun. That extrapolation cannot be verified by examining the actual photometric variability of solar-like stars, because no

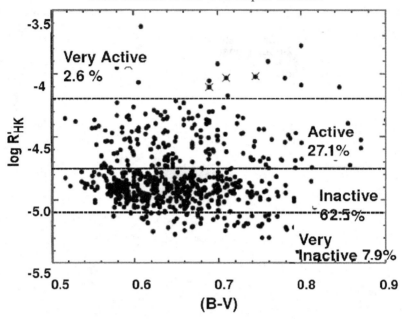

FIGURE 4. Activity indicator $R'_{HK}$ versus stellar spectral type (Henry et al. 1966). The Sun lies in the middle of the inactive region. Its value of $R'_{HK}$ varies throughout the bin during the solar activity cycle. Note that most stars are less active than the Sun.

star other than the Sun has been measured to the requisite precision. However, the $R'_{HK}$ index is believed to be well correlated to stellar variability. It is based on the spectral line profile of the Calcium II H and K lines and is readily measured with ground-based telescopes.

Figure 4 shows measurements of the $R'_{HK}$ index for a variety of spectral types. As can be seen from this figure, about 70% of the stars are found to have an index at least as low as that of the Sun. Hence we plan to choose approximately 150,000 late-type dwarfs to monitor during the first year of observations, and then gradually eliminate those that are too variable, to find Earth-size planets. This action will limit the time needed by the Deep Space Network to receive telemetry from the spacecraft as it recedes from Earth.

Although most solar-like stars are expected to have stellar activity levels no higher than that of the Sun, planets can still be found around stars with higher activity levels if the size of the planets are somewhat larger than Earth, or if:

- They are found around later spectral types;
- The stars are brighter (less shot noise);
- They are closer to the central star (more transits);
- The transit is closer to a central transit than assumed here (grazing transit).

Planets in the HZ of K dwarfs have orbital periods of a few months, and therefore would show about 16 transits during a four-year mission. Figure 5 shows the minimum size a planet would require to produce an $8\sigma$ detection versus the amplitude of the stellar variability, assuming that the frequency distribution of the stellar noise is the same as that of the Sun. The upper curve shows that the amplitude of the stellar noise would need to be at least eight times that of the Sun before it would prevent planets slightly larger than twice the radius of the Earth from being detected. For short-period planets showing 16 transits, planets as small as 1.4 times the radius of the Earth would still be

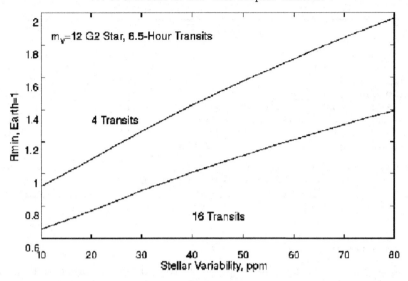

FIGURE 5. Effect of increased stellar variability on the minimum-sized planet that can be detected with $8\sigma$.

detectable for stars, even if the star had eight times the amplitude variability of the Sun (see also Jenkins 2002).

## 6. Interplay of noise from stellar variability, Poisson variations, and the instrument

As described earlier, it is important to find planets with a size sufficient to produce $8\sigma$ detections with three or more transits. Equation (6.1) presents a simplified relationship of the SNR as a function of ratio of the area of the planet to area of the star $(A_p/A_*)$, number of transits $(N_{\mathrm{tran}})$, and the noise due to stellar variability $(v)$, Poisson noise in the stellar flux $(F)$, and instrument noise $(i)$:

$$\mathrm{SNR}\ \alpha (N_{\mathrm{tran}})^{1/2} \frac{(A_p/A_*)}{[(i^2 + v^2) + 1/F]^{1/2}}\ , \tag{6.1}$$

where $i$, $v$, and $F$ are in electrons counted over a sample period equal to the transit duration. It should be noted that $i$ represents many separate noise sources (Koch 2002), and that noise variance due to stellar variability is a function of the transit duration (see Figure 3). An examination of Equation (6.1) shows that for very bright stars where the shot noise and instrument noise are small, the SNR is dominated by the stellar variability. The opposite is true for dim stars where $1/F$ is large and the SNR is dominated by the shot noise in the stellar flux. Table 1 presents calculated noise values for *Kepler*. For all entries in this table, a stellar variability of 10 ppm is assumed. Note that at 14th magnitude, the effect of stellar variability and instrument noise are negligible, and the photometric precision is dominated by the shot noise. It is also clear that for stars 14th magnitude and dimmer, detectable planets must be somewhat larger than the diameter of the Earth if they are to be reliably detected orbiting G2 dwarfs. Planets much smaller than the Earth are readily found if they orbit smaller stars, are in short-period orbits, or orbit the 70% of stars that are expected to have lower variability than the Sun.

| | 9 | 12 | 14 |
|---|---|---|---|
| Visual magnitude | 9 | 12 | 14 |
| Stellar signal (photo electrons) | $6.5 \times 10^{10}$ | $4.1 \times 10^9$ | $6.5 \times 10^8$ |
| Stellar shot noise (ppm) | 3.9 | 16 | 39 |
| Instrument noise (ppm) | 2.6 | 6.6 | 12 |
| Solar variability (ppm) | 10 | 10 | 10 |
| Combined differential photometric precision (ppm) | 12 | 20 | 42 |
| Relative signal for Earth transit across the Sun (ppm) | 84 | 84 | 84 |
| SNR for four transits | 14 | 8.4 | 4.0 |
| Minimum detectable planet radius (Earth = 1) at $8\sigma$ | 0.75 | 0.98 | 1.4 |

TABLE 1. Comparison of SNR and minimum size planet that produces an $8\sigma$ detection for a planet transiting a G2 dwarf

## 7. Expected results

As the mission duration lengthens from months to years, detection and follow-up begin with the detection of three or more transits for larger planets in short-period orbits, and then expands outward to the detection of smaller planets and longer orbital periods. For terrestrial-size planets with orbital periods of a week or less and geometrical alignment probabilities of 10%, the number of discoveries during the first 30 days of observations should be of order: (100,000 stars * 10% alignment probability * fraction of stars with such planets). The predicted number of discoveries varies from 10,000 to 100 as the fraction of stars with planets in inner orbits varies from 1 to 0.01.

To estimate the number of planet discoveries expected as the mission progresses, a model of a planetary system was convolved with both the distributions of stars in the FOV and the system response. To assess the level of resources needed to examine the expected number of candidates, the planetary model makes the optimistic assumption that each star has a planetary system with two Earth-size planets positioned outside the HZ of the star, and that there is one terrestrial-size planet in its HZ. It is assumed that planets found outside the HZ could be present with semi-major axes similar to those already found for the 150 giant planets already discovered. The model assumes that there is an equal probability of finding one of the two planets at 0.05, 0.1, 02, 0.4, 0.6, 0.8, 1.0, 1.2, or 1.5 AU. Whenever such a planet falls in or near the HZ of a star, it is removed to avoid conflict with the terrestrial-size planet already assumed to be in that position. The results shown in the figure are readily scaled to other assumptions and situations.

It is clear from Figure 6 that nearly 1300 Earth-size planets in short-period orbits will be found during the first year of operation and that approximately 4,500 will be found after four years of operation. Most of this contribution occurs for inner planets because they have a high probability of geometric alignment, and because the short-period orbits produce a large number of transits that greatly increases the SNR of the transit pattern. Although these values are probably high, they are useful for obtaining a conservative estimate of the magnitude of the analysis and follow-up observation tasks. It should also be noted that if planets larger than Earth size are common, they will produce higher SNR than assumed here, will be more easily detected around the many dim stars observed in the *Kepler* survey, and will thereby increase the values shown in Figure 6. Over the duration of a four-year mission, the number of transits in a pattern will exceed 400 for the shortest period orbits. The resulting increase in the SNR of the folded data will exceed eight even for planets as small as Mars or Mercury. Consequently, *Kepler* should produce an excellent estimate of the size distribution of terrestrial planets.

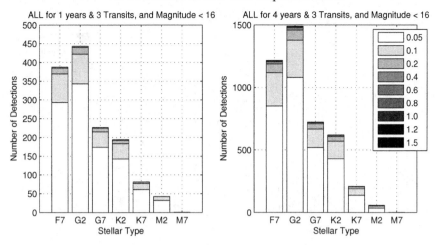

FIGURE 6. The number of Earth-size planets expected to be discovered as a function of stellar spectral type and semi-major axis and mission duration.

The model treats planets in the HZ differently than those placed outside it. Here, the objective is to determine the smallest possible planet that provides an SNR greater than the $7\sigma$ threshold value. The diameter of the terrestrial-size planet placed in the HZ of each star is systematically varied between values of Mars to 1.5 times the radius of Earth (0.53, 1.0, 1.3, 2.0 $R_\oplus$) and the smallest detectable planet size is tabulated. The capability to detect small planets is valuable, because a non-detection rules out a larger range of planet sizes, and their detection provides information on the tail of the distribution of planet sizes. Figure 7 shows how many of each size can be detected for both four- and six-year missions and for stellar types from F7 to M7. Figure 7 shows that for a mission duration of four years, no terrestrial planets can be found for stars as early as F7. This result is dictated by the requirement to have at least three transits for a valid discovery, and because planets in the HZ of early spectral types have orbital periods exceeding 1.5 years. However, when the mission duration is extended to six years, over 100 terrestrial-size planets can be discovered. A comparison of the results for a four-year versus six-year mission also shows that total number of expected discoveries increases, and that the fraction of the total discoveries that are capable of detecting the smallest planets rises. Because of the small sizes of the M dwarfs, and because planets in their HZ provide many transits per year, planets as small as Earth and Mars can be detected. Detection of a total 650 terrestrial-size planets, including approximately 100 Earth-size planets, is expected.

With the ability to discover about 650 terrestrial planets in the HZ for a four-year mission and about 900 for a six-year mission, our knowledge of terrestrial planets of a wide variety of stellar types in the HZ may exceed that of extrasolar giant planets. Even if only 10% of the stars have such planets, correlations can be obtained for the frequency of planets with stellar type and metallicity. Given that several percent of solar-like stars have orbiting giant planets, it seems unlikely that as few as 1% of such stars have terrestrial planets. Nevertheless, if that should be the case, *Kepler* will find hundreds of Earth-size planets in inner orbits and at least a few terrestrial-size planets in the HZ. If no terrestrial-size planets are found, then they must be very rare and theories on terrestrial planet accretion and migration will need to be re-examined.

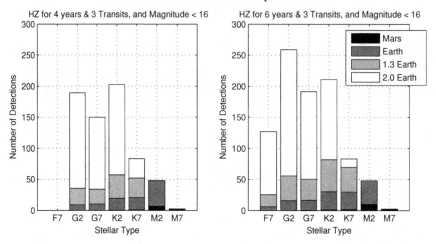

FIGURE 7. Number of terrestrial-size planets expected to be discovered as a function of stellar spectral type, semi-major axis, and planet size. It is assumed here that each target star has one terrestrial-size planet in its HZ.

## 8. Validation of planet detections

Before a candidate detection can be considered to be a validated planet, a rigorous validation process must be executed to ensure that it is not due to some other phenomenon (Borucki et al. 2003). Public release of false positives would ultimately discredit any mission results. Therefore, to be considered a validated planet, the detection must meet several requirements:

1. The total statistical significance (SNR) of the superimposed transits must exceed $7\sigma$. This requirement prevents false positives produced by statistical noise when $8 \times 10^{11}$ statistical tests are carried out on $10^5$ stars for orbital periods from 1–700 days.

2. At least three transits must be observed that demonstrate a period constant to 10 ppm. This test is independent of the previous test and demonstrates the presence of a highly periodic process. It essentially rules out mistaking stellar phenomenon for transits. (Exceptions must be made for planets showing timing variations caused by mutual perturbations.)

3. The duration, depth, and shape of the light curve must be consistent. The duration must be constant over all transits and consistent with Kepler's laws based on the orbital period. The depth must be consistent over all transits. A weaker requirement is that the shape must be consistent with a "U" shape of a planetary transit, rather than a "V" shape of a grazing eclipse of a binary star. Low-amplitude transits are likely to be too noisy to make this distinction.

4. The position of the centroid of the target star determined outside of the transits must be the same as that of the differential transit signal. If there is a significant change in position, the cause of the signal is likely to be an eclipsing star in the background.

5. Radial velocity measurements must be conducted to demonstrate that the target star is not an eclipsing binary with the period of the transits.

6. High-precision radial velocity measurements must be made to measure the mass of the companion or provide an upper limit that is consistent with that of a small planet.

7. High spatial-resolution measurements must be made of the area immediately surrounding the target star to demonstrate that there is no background star in the aperture capable of producing a false positive signal.

Requirements #1 and #2 insure that statistical fluctuations in the data series will produce less than one false positive for the entire mission, which is critical for the case of a null result. Requirements #3 through #7 greatly reduce the probability that unrelated physical phenomena could be mistaken for a pattern of planetary transits.

Other checks are also possible. For example, if future instrumentation on *HST* and *JWST* has sufficient precision to detect the color changes during the transit, a measured color change consistent with the differential limb darkening expected of the target star would strengthen the validation (Borucki & Summers 1984). A shape, depth, or color change substantially different than expected would point to the possibility of a very close background star that differed in spectral type or reddening.

Giant planets like 51 Pegasi, with orbits of less than seven days, are also detected by the periodic phase modulation of their reflected light without requiring a transit (Borucki et al. 1997). For the short-period giant planets that do transit, the planetary albedo can be calculated. Information on the scattering properties of the planet's atmosphere can also be derived from the phase curves (Marley et al. 1999; Seager et al. 2000; Sudarsky et al. 2000).

Ground-based Doppler spectroscopy and/or space-based astrometry with SIM can be used to measure the larger planetary masses and can distinguish between a planet and a brown dwarf. These complementary methods can also detect additional massive companions in the systems to better define the structure of each planetary system. The density of any giant planet detected by both photometry and either radial velocity or astrometry can be calculated. Determination of the planet size, mass, semi-major axis, and stellar properties provides the properties critical for the validation and development of theoretical models of planetary structure.

White dwarf stars are about the size of the Earth and might be expected to produce a transit signal of similar magnitude. However, because of the gravitational lensing caused by their large, but compact, mass, the transits actually result in an increase in brightness (Sahu & Gilliland 2003) and are thereby readily distinguished from those of a planet.

## 9. Data release policy

To avoid claims based on false positive detections, the series of data validation procedures and ground-based observations must be made and analyzed to validate the discovery before any announcements are made. It is expected that several months will be required to obtain telescope time on suitable ground-based telescopes and to conduct the observations and analysis. The limited seasonal availability of the FOV and effects of poor weather will add to the delay. Hence, the data can not be released until several months have passed from the time that the detection algorithm shows that at least three transits with a combined SNR > $7\sigma$ have been detected.

When data are released, both original calibrated data for each target and the light curves generated by ensemble photometry will be provided so that all interested members of the scientific community can independently assess the reliability of the results.

## 10. Education and public outreach programs

Bringing the science and the excitement of space missions to students and the general public is a major goal of all NASA efforts. To further this goal, the *Kepler* mission includes two institutions that produce high quality educational materials and have high impact in education on a national level: the Lawrence Hall of Science (University of California) and the SETI Institute. Formal education contributions will include Great Explorations in Math and Science (GEMS) teacher guides in math and science for grades K through 8, Full Option Science System (FOSS) teacher workshops for middle schools, and the addition of a *Kepler*-science module in the Hands-On Universe high school curriculum. For informal science education, *Kepler* has developed an orrery demonstrating the detection-by-transit method. This "hands on" device is now part of a traveling museum exhibit sponsored by the Space Science Institute (SSI). The informal education program is also developing programs explaining planet finding for small and medium-size planetariums serving both school and public audiences. *Kepler* will fund the creation of public radio broadcasts through *Stardate* and a video program suitable for public broadcast. To enhance participation by the public, predictions of the large-amplitude transits by giant planets will be distributed via the *Kepler* website (http://Kepler.NASA.gov), so that amateur astronomers and institutions with CCD-equipped facilities can observe planetary transits. To increase the use of CCD technology, several CCD cameras will be supplied to minority colleges that already have telescopes. Training and support will be provided to the college faculty in photometric observing techniques and data analysis methods required for high precision photometry in discovery and observation of planetary transits.

## 11. Opportunities to participate

It is expected that several opportunities to participate in the *Kepler* mission will be available. Current concepts envision a Participating Scientist Program (PSP) with a Guest Observer (GO) option and a Data Analysis Program (DAP).

PSP investigators will be encouraged to propose research topics that complement those of the Mission Team. It is expected that the PSP will have two options: the Participating Scientist (PS) option, whereby the participating scientist proposes a research program directly concerned with detection, characterization, or understanding of extrasolar planets; and the GO option, whereby the proposer specifies targets in the *Kepler* FOV that are to be observed for astrophysical interest, but would not otherwise be included on the target list.

For the PS option, the proposed programs can be analytical, observational, or theoretical in nature. Examples of appropriate analytic programs include: modeling eclipsing binary systems to determine the characteristics of the stars and planets, and measuring and modeling timing variations in the epoch of transits to detect non-transiting planets. The PS option can involve use of the existing target data, or requesting observations of additional targets. Examples of ground-based observational programs include those designed to fully characterize stars found to have planets, high-precision radial velocity measurements to determine the mass of the planets detected from transits or from reflected light, confirmation of transits or efforts to detect atmospheric absorption, and observations that verify that the transit signal is coming from the target star rather than a background star.

The GO option should accommodate those investigators who wish to make astrophysical measurements of the many different types of objects in the *Kepler* FOV. Generally, these targets will be different than those chosen for the transit search. Examples include variable stars of all types, distribution and time variation of zodiacal light, and extragalactic objects. It is expected that, at any one time, a total of about 3,000 additional targets will be available, and that these selections can be changed at intervals of three months. Most of the targets will be observed at a cadence of once per 15 minutes, but a small subset could be observed with a one-minute cadence. All targets must be within the active area of the *Kepler* FOV. The FOV will not be moved to accommodate a GO request. For stars already on the *Kepler* target list, the GO will be referred to the DAP.

The *Kepler* target list is expected to be available at launch to allow investigators to plan their requests.

Investigators desiring to analyze data from targets already on the *Kepler* target list will apply to the Data Analysis Program (DAP). DAP is an opportunity for the scientific community to perform data mining on the existing database. Examples of potential uses for the data are validation of planetary detections, exoplanet searches using alternative techniques, analysis of stellar activity cycles, white-light flaring, frequency of Maunder minimums, distribution of stellar rotation rates, etc.

A data release policy has been developed to release Mission data at the earliest time that allows for data calibration and validation and insures against false-positive planetary detections. The first three-month data set will be released approximately one year after commissioning, and then supplements will be released annually. Publicly released *Kepler* observations will be freely available to all interested parties for data mining. The data will be archived in the Multimission Archive at Space Telescope (MAST) and supported for at least five years after the end of the mission.

## REFERENCES

BORUCKI, W. J., KOCH, D., BOSS, A., DUNHAM, E., DUPREE, A., GEARY, J., GILLILAND, R., HOWELL, S., JENKINS, J., KONDO, Y., LATHAM, D., & REITSEMA, H. 2004. In *Second Eddington Workshop: Stellar Structure and Habitable Planet Finding* (eds. F. Favata, S. Aigrain & A. Wilson). p. 177. ESA.

BORUCKI, W. J., KOCH, D. G., DUNHAM, E. W., & JENKINS, J. M. 1997. In *Planets Beyond the Solar System and the Next Generation of Space Missions* (ed. D. Soderblom). ASP Conference Series Vol. 119, p. 153. ASP.

BORUCKI, W. J. & SUMMERS, A. L. 1984 *Icarus* **58**, 121.

HENRY, T. J., SODERBLOM, D. R., DONAHUE, R. A., & BALIUNAS, S. L. 1966 *AJ* **111**, 439.

JENKINS, J. M. 2002 *ApJ* **575**, 493.

JENKINS, J. M., CALDWELL, D. A., & BORUCKI, W. J. 2002 *ApJ* **564**, 495

KOCH, D. G. 2004. In *Bioastronomy 2002, Life Among the Stars* (eds. R. Norris & F. Stootman). IAU Symp. 213, p. 85. ASP.

KOCH, D. G., ET AL. 2006. In *A Decade of Extrasolar Planets around Normal Stars: Poster Papers from the Space Telescope Science Institute Symposium, May 2005.* p. 21. STScI.

LINEWEAVER, C. H. & GRETHER, D. 2003 *ApJ* **598**, 1350.

MARLEY, M. S., GELINO, C., STEPHENS, D., LUNINE, J. I., & FREEDMAN, R. 1999 *ApJ* **513**, 879.

RIVERA, E. J., LISSAUER, J. J., BUTLER, R., MARCY, G. W., VOGT, S. S., FISCHER, D. A., BROWN, T. M., LAUGHLIN, G., & HENRY, G. W. 2005 *ApJ* **634**, 625.

SAHU, K. C. & GILLILAND, R. L. 2003. *ApJ* **584**, 1042.

SEAGER, S., WHITNEY, B. A., & SASSELOV, D. D. 2000 *ApJ* **540**, 504.

SUDARSKY, D., BURROWS, A., & PINTO, P. 2000 *ApJ* **538**, 885.

# Observations of the atmospheres of extrasolar planets

By TIMOTHY M. BROWN,[1] ROI ALONSO,[2]
MICHAEL KNÖLKER,[1] HEIKE RAUER,[3]
AND WOLFGANG SCHMIDT[4]

[1]High Altitude Observatory, National Center for Atmospheric Research,† P.O. Box 3000,
Boulder, CO 80307, USA

[2]Instituto de Astrofisica de Canarias, La Laguna, Tenerife, SPAIN

[3]Institute of Planetary Research, DLR, Rutherfordstrasse 2, 12489 Berlin, Germany

[4]Kiepenheuer Institut für Sonnenphysik, Freiburg, Germany

The extrasolar planets known to date have masses and orbital periods spanning a large range. Those for which we have definite knowledge about physical composition have much more restricted properties: they are either transiting planets with near-Jovian masses and orbital periods of a few days, or (as in a couple of recent discoveries) they are distant low-mass companions to objects that are themselves low-mass and young. Here we will concentrate on the former group of objects, and try to summarize what is known and conjectured concerning their atmospheres based on observations of their transits. By way of motivation and illustration of the ultimate possibilities available to transit observations, we begin by discussing recent observations of the transit of Venus in June 2004.

---

## 1. Introduction

Since the discovery, a decade ago, of the first planet orbiting a Sun-like star (Mayor & Queloz 1995), radial velocity surveys have taught us much about the nature of other solar systems. In particular, the distribution of masses, orbital semi-major axes, eccentricities, and parent star metallicities have driven home the importance of migration processes in protoplanetary disks, and strongly suggested an important role for metallicity in planet formation (e.g., Marcy et al. 2003; Santos et al. 2003; Thommes & Lissauer 2005; Fischer & Valenti 2005). Similarly, the relative abundance of multi-planet systems (with components often lying in resonant orbits) raises intriguing questions about the interactions between larger bodies in such circumstances (e.g., Kley et al. 2005; Laughlin et al. 2005).

But radial velocity measurements alone do not provide any direct information about the planets' composition, structure, or internal dynamics. To learn these things, we need to spatially resolve the planet from its host star so that we can perform spectroscopy on it alone, or we need some surrogate of this process. A useful surrogate is available for the small fraction of extrasolar planets whose orbits are aligned so that we can see them transit the disks of their parent stars. During such primary transits, when the planet passes between us and the star, the overwhelming light from the parent star can be used to the observer's advantage; it allows high S/N measurements to be made so that the geometry of the transit can be tightly specified, and so that small differences between in- and out-of-transit spectra can be detected. During secondary transits, the planet passes from sight behind the star within the span of a dozen or so minutes; this rapid disappearance allows accurate time-differential flux measurements to be performed, giving the ratio of fluxes between the planet and the star.

† The National Center for Atmospheric Research is supported by the National Science Foundation

These methods are beginning to answer some of the basic questions concerning the hot, close-in extrasolar planets. Are they giant, dense, red-hot balls of nickel and iron? (No.) Are they round, or do they have square corners? (Round planets fit the data better.) Do they have clouds, and weather? (Very likely, though details are obscure.) In what follows, we will try to explain what we think we know about such things, and how we know it.

## 2. Basic considerations

If an opaque spherical planet were to pass across a uniformly illuminated stellar disk, the fractional decrease in light arriving from the star $\delta I/I$ would be simply

$$\frac{\delta I}{I} = -\frac{R_p^2}{R_*^2} \ , \qquad (2.1)$$

where $R_p$ is the radius of the planet and $R_*$ is that of the star. The lowest-order information from a stellar transit is thus the ratio between the planetary and stellar cross-sectional areas.

Two useful complications arise right away. The first is that stellar disks are not uniformly bright; rather, they are limb-darkened (at visible and near-IR wavelengths, anyway), with the amount and functional form of the limb darkening depending on the wavelength of observation. Thus, even a simple-seeming measurement of the ratio of radii is unreliable (at the 20% level) unless we know the transit's impact parameter, i.e., the minimum separation between the stellar and planetary centers, projected on the plane of the sky. For circular orbits, this is the same as knowing the orbit's semi-major axis $a$ and inclination to our line of sight $i$. Fortunately, the orbital inclination has other effects on the observed light curve (decreasing $i$ from 90° shortens the total duration of the transit and lengthens the times required for ingress and egress when the planet's disk lies only partly on that of the star). The upshot is that one can characterize a transit light curve with four parameters as illustrated in Figure 1, two of which ($l$ and $d$) are measureable in practice with relative ease and accuracy, and two of which ($w$ and $c$) are more difficult and error prone. Given the orbital period, an estimate of the stellar mass, and these four observables, it is possible to make separate estimates of $R_*$, $R_p$, $i$, and a limb-darkening parameter $c_l$. If one is willing to stipulate the value of any of these parameters ($c_l$, say, based upon stellar atmospheres models), then the others can be determined more accurately, or with poorer data. These considerations (combined with mass estimates from radial velocities) have led to our present knowledge about the mean densities of extrasolar planets, which we shall discuss briefly below.

The second complication is that, for purposes of a transit light curve, $R_p$ is the radius at which the planet becomes opaque to tangential light rays at the wavelength of observation. This is generally not the same as the radius of the photosphere (which relates to light rays propagating more or less vertically, and wavelength averaged to boot), and it is certainly not the same as some arbitrary (e.g., one bar) pressure level in the atmosphere (Burrows et al. 2003). The dependence on pressure level is important for comparisons with self-consistent models of the planetary structure. The wavelength dependence, however, provides diagnostics for the composition, dynamics, and thermodynamic state of the planetary atmospheres.

Consider first a "continuum" wavelength $\lambda_c$, at which the atmosphere is relatively transparent, and let the zero point for measuring height in the atmosphere be that point at which the optical depth $\tau_c$ along a tangential ray is unity. If we now choose a nearby wavelength where the opacity $\kappa_l$ is larger (say, the center of a strong molecular absorption line), then unit tangential optical depth at this wavelength will occur at a greater height,

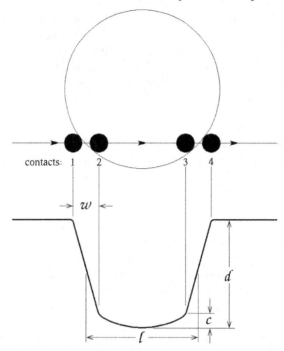

FIGURE 1. The light curve of a planet transiting the disk of a star, showing four parameters (transit depth, total duration, ingress/egress time, and bottom curvature) that can be measured with photometry of suitable quality.

such that the decrease in mass density compensates for the increase in opacity per gram. If the atmosphere is locally exponentially stratified with a density scale height $H$, and the absorbing species is well mixed, then one can express the necessary height change $\delta z$ as

$$\delta z = H \ln \left( \frac{\kappa_l}{\kappa_c} \right) \quad . \tag{2.2}$$

Thus, to compensate for an increase in opacity by a factor of $e$, one must move upward in the atmosphere by one scale height. This means that the apparant radius of the planet depends upon wavelength, and so too does the fraction of stellar light blocked by the planet during a transit.

How large is this effect? In principle, it can be substantial. The opacity ratio between strong lines and continuum for relevant molecular vibration-rotation lines is pretty often $10^4$, corresponding to 10 atmospheric scale heights. Moreover, the atmospheres of close-in giant planets likely have low molecular weight, high temperature, and modest surface gravity. As a result, $H$ can exceed 500 km, which may be more than 0.5% of the planetary radius. Thus, according to this simple calculation, small changes in wavelength might change the total light obstructed by the planet by as much as 10%. For further details, see Seager & Sasselov (2000) and Brown (2001). The wavelength-dependent variations seen so far are not as large as this, but they are, nevertheless, big enough to detect in at least one case.

Another interesting situation is that of a cloud deck composed of large (larger than $\lambda$) particles, distributed in a layer with a well-defined top at height $z_p$. Such a cloud deck will indiscriminately scatter or absorb tangential rays of all wavelengths. The planet will then appear to have the same radius at all wavelengths, except those for which the cloud-free continuous opacity is so large that the tangential optical depth is greater than unity at $z_p$. High clouds can, therefore, suppress the atomic and molecular absorption signatures that one might otherwise see.

Finally, consider thermal emission from the planet. At visible wavelengths, most of the light emitted from the planet's day side is expected to be reflected starlight. The maximum possible reflected signal relative to the stellar flux is simply $(I_p/I_*) = \pi R_p^2/(4\pi a^2)$, which for close-in giant planets typically gives $(I_p/I_*) \simeq 3 \times 10^{-5}$. If the planetary albedo is less than unity, this number is, of course, even smaller. In the thermal infrared, however, things are quite different. For the expected planetary equilibrium temperatures of roughly 1000 K, the peak of the Planck function falls at about 3.5 $\mu$m wavelength. Wavelengths in the 4–25 $\mu$ range, observable by the *Spitzer Space Telescope*, thus begin to approach the Rayleigh-Jeans long-wavelength limit, in which $I_p/I_* = (R_p^2/R_*^2)(T_p/T_*)$, where $T_p$ and $T_*$ are the effective temperatures of the planet and star, respectively. For most known transiting extrasolar planets, this intensity ratio is a few times $10^{-3}$, a much more accessible signal than in visible light. This, then is the motivation behind recent successful attempts to measure the diminution in IR flux during the secondary transits of HD 209458b and TrES-1b.

## 3. Transit of Venus

Twice (or rarely, three times) every 120-odd years, Venus passes across the solar disk as seen from the Earth. Such an event occurred in June 2004, providing the first opportunity to observe the transit of a planet with an atmosphere across the Sun since the advent of quantitative spectroscopy. This was thus an opportunity to demonstrate in a graphic way some of the techniques that we would like to appy to much more distant planets, and in addition, it offered the possibility of learning something new about Venus's atmosphere. Accordingly, we undertook observations of the June 2004 transit, using the Vacuum Tower Telescope (VTT) of the Kiepenheuer Institute for Solar Physics (KIS) on Tenerife.

Venus's atmosphere consists mostly of $CO_2$, which has strong vibration-rotation absorption bands in the near IR; we chose to observe at a small number of wavelengths near 1.5 $\mu$m, where in a short wavelength range one can find lines that become opaque to tangential rays at all heights between the cloud tops (at about 65 km above the solid surface) and the mesopause near 110 km.

A great advantage of the Sun's proximity is that it can be spatially resolved, so that one can ignore the vast majority of its light that does not pass near Venus's limb. The geometry of the transit observations is illustrated in Figure 2. We placed the slit of the KIS spectrograph at a fixed place on the Sun, and allowed Venus's motion to carry it across the slit. The pixel size and slit width were both about 0.4 arcsec, or roughly 80 km projected onto Venus's disk. We used the VTT adaptive optics system, which held the effective seeing width to about 1 arcsec throughout the five hours occupied by the transit. The spectral resolution was about $R = \lambda/\delta\lambda = 200{,}000$. A consequence of the high spectral resolution and the small (256 × 256 pixels) detector format was that only a small wavelength range could be observed at one time. Since there was some uncertainty in our estimations of $CO_2$ line strengths, we took observations in four neighboring wavelength regions with varying combinations of line strengths.

At any instant we recorded a spectrum such as that shown in Figure 3. At either end of the slit one sees only the solar spectrum, modified by absorption by $CO_2$ in the Earth's atmosphere. Where Venus's disk crosses the slit, the observed intensity is much smaller; it is not essentially zero only because of scattered light in the Earth's atmosphere and in the telescope and spectrograph optics. The excitement occurs where Venus's limb crosses

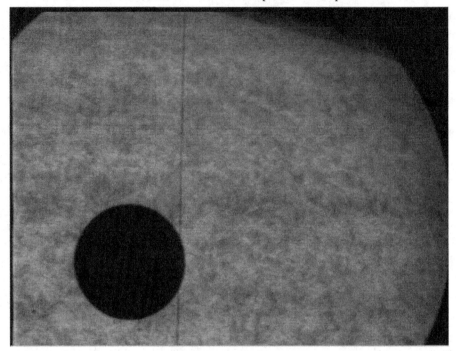

FIGURE 2. Slit-jaw image of Venus on the solar disk as observed at the KIS vacuum tower telescope on Tenerife, showing the vertical spectrograph slit nearly tangent to Venus's disk. The prominent circular boundary on the right-hand side is the edge of the adaptive-optics aperture, while the limb of the much larger solar disk cuts diagonally through the top of the image.

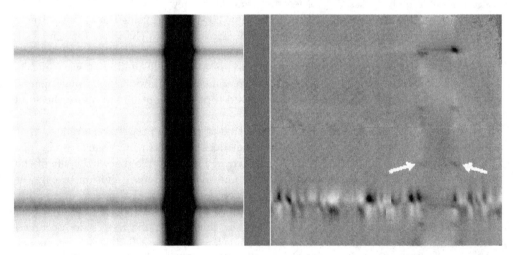

FIGURE 3. (Left panel) Typical spectrum of the Sun and Venus, with distance along the slit $x$ increasing from left to right, and wavelength $\lambda$ increasing from bottom to top. (Right panel) The difference between the spectrum at left and a simple model of it as the product of a function of $x$ and a function of $\lambda$, as described in the text.

the spectrograph slit—here a portion of the observed light has passed through Venus's atmosphere as well as our own. By comparing the spectra from these regions to those of unadulterated sunlight, one can infer Venus's contribution to the absorption.

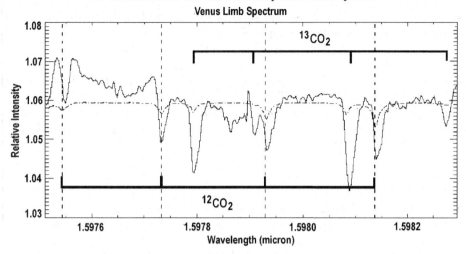

FIGURE 4. Transmission spectrum of Venus's atmosphere, averaged over a few hundred spectra taken at different latitudes. The dot-dashed line shows the absorption spectrum of the Earth's atmosphere from Livingston & Wallace (1991), on an arbitrary vertical scale. Lines resulting from $^{12}CO_2$ and $^{13}CO_2$ are identified.

The spectrum in the left panel of Figure 3 shows two strong spectrum lines; both are $CO_2$ lines, but the lower one is underlain by an atomic line originating in the solar photosphere. The right panel of the figure shows a difference spectrum, constructed from the left-hand image as follows: We created a model of the observed spectrum by forming averages along both $\lambda$ and $x$, and then forming the two-dimensional product of these two one-dimensional marginal functions. We then subtracted a suitably scaled version of this model image from the observed one; the right-hand panel displays this difference.

The difference image reveals a fringe structure that originates in the instrument; at present our ability to retrieve scientific information from these images is limited by our ability to model this fringing successfully. The purely telluric absorption line almost disappears from the difference image, showing that it is accurately represented as the product of an average line profile and the average intensity along the slit. The solar line shows significant residuals, however, arising from local variations in velocity and intensity in the solar photosphere. These are caused mostly by the solar granulation. Finally, near Venus's limb, one can see localized dark spots (two are indicated by arrows). These are the signature of absorption by Venus's atmosphere.

We have constructed an average spectrum of the Venusian absorption by averaging together several hundred individual spectra like that shown in Figure 3, with due allowance for the changing position of Venus's limbs on different spectra. The result for one of our four wavelength bands (a different one than illustrated in Figure 3) is shown in Figure 4. Also shown in the figure is a spectrum of the telluric absorption as seen at one airmass, taken from the high-resolution IR spectral atlas by Livingston & Wallace (1991).

As indicated in the figure, the observed lines can be identified with two different isotopomers, namely $^{12}CO_2$ and $^{13}CO_2$. A curious feature of the spectrum is that the strengths of the $^{13}C$ lines are similar to or greater than those of the $^{12}C$ lines, even though $^{12}C$ is by far the more abundant isotope. In fact, some of the $^{13}CO_2$ lines in the Venusian spectrum are significantly stronger, relative to the $^{12}CO_2$ lines, than in the telluric spectrum. This behavior occurs because the lower levels of the two sets of lines are dissimilar in energy: for the $^{13}CO_2$ lines the lower level lies 200 cm$^{-1}$ above the ground energy, while for the $^{12}CO_2$ lines it is more than 700 cm$^{-1}$ higher. At the cold temperatures prevailing near Venus's mesopause (about 200 K), the pile-up of molecules in lower-lying states is enough to compensate for the low relative abundance of $^{13}CO_2$.

Work is continuing to resolve issues related to the detector fringing, in hopes of measuring accurate Doppler shifts of the lines seen in Figure 4. If this can be done, it will be possible to measure the cross-terminator wind speed as a function of Venusian latitude and (by measuring lines of differing strength) height above the cloud deck. Such measurements could be very useful for understanding the way in which Venus's atmosphere makes the transition from super-rotation (near the top of the cloud deck) to the dayside-to-nightside flow that dominates at much higher altitudes.

## 4. Primary transits

The primary transit by an extrasolar planet, in which the planet passes between its parent star and the Earth, is useful for several reasons. The light curves from such transits have been used to estimate gross properties of all of the presently known transiting planets and of their stars, measuring the stellar and planetary radii, the inclination of the planets' orbit, and the strength of limb darkening on the stars. Moreover, for the two nearest transiting planets, more-or-less successful attempts have been made to use the variations with wavelength in the depth of the observed transits to learn about the composition of the planets' atmospheres. I shall further discuss all of these results below. In addition to these measurements, one can use radial velocity measurements taken during the primary transit to infer the inclination of the planetary orbital axis to the rotational axis of its star (e.g., Winn et al. 2005). Although this procedure (based on the Rossiter-McLaughlin effect) is very interesting, I shall not discuss it further here.

### 4.1. *Bulk properties*

As described in the introduction, the observed transit light curves may be parameterized (conceptually, at least) in terms of four quantities: the transit duration, its depth (measured relative to the out-of-transit intensity of the star), the duration of the ingress and egress phases of the transit, and the curvature of the light curve between the end of ingress and the beginning of egress. This parameterization is useful for achieving an intuitive understanding of how light curve properties relate to those of the star/planet system, and, in fact, it is the basis for a helpful technique for distinguishing between true planetary transits and false alarms (which merely resemble planetary transits) involving grazing eclipses or multiple-star systems (Seager & Mallén-Ornelas 2003).

To obtain numerical estimates of planetary properties, however, one usually bypasses the approximations inherent in the four-element parameterization. Rather, one performs a suitable error-weighted fit to the photometric and radial velocity observations, adjusting the parameters of a detailed numerical model so as to minimize the mismatch with observed data. The most significant parameters to emerge from these fits are, of course, the planetary masses and radii. The relation between them is shown in Figure 5, for all of the transiting planets known as of this writing (1 Sept. 2005). This figure is a compilation of results from Brown et al. (2001); Udalski et al. (2002a,b); Konacki et al. (2003, 2005); Moutou et al. (2004); Alonso et al. (2004); Sozzetti et al. (2004); Pont et al. (2004), and Sato et al. (2005), with photometry and radial velocities coming from a wide variety of sources. There is a fairly clear distinction between the OGLE planets (which, because they are faint, show relatively large uncertainties in both mass and radius) and the planets of brighter stars (HD 209458b, TrES-1, and HD 149026b, for which the uncertainties are smaller). The radius uncertainties are not, however, tremendously different between the two sets of stars. The reason is that, beyond a certain level of precision, uncertainties

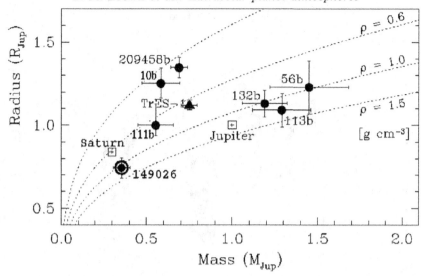

FIGURE 5. Observed mass/radius relation for transiting extrasolar planets. The observed mean densities are mostly consistent with hydrogen/helium gas giants that lack significant internal energy sources. Exceptions are HD 209458b, which is less dense than predicted for such objects, and HD 149026b, which is more dense.

in the planetary radius are dominated by uncertainties in the radius of the parent star. With superb photometry (as is obtainable with *HST*, for instance) it is possible to estimate the stellar and planetary radii separately, but the stellar radius estimate remains fairly imprecise. Thus, in these cases the ratio $R_p/R_*$ is much better known than either individual value.

Most of the planets shown in Figure 5 have radii that are in fairly close agreement with theoretical predictions—assuming that the planets have near-solar composition, migrated near to their stars early in their lives, and contain no significant internal energy sources (e.g., Bodenheimer et al. 2003; Burrows et al. 2004; Chabrier et al. 2004, but also see Gaudi 2005). Exceptions are HD 209458b (and possibly OGLE-Tr-10b), which are larger than theory predicts, and HD 149026b, which is smaller. The latter object has roughly Saturn's mass, but considerably higher density, in spite of orbiting very close to its star. It is possible to explain this high density by assuming that the planet contains a very large core of material with high atomic weight (60–80 Earth masses, as compared to a total planetary mass of only about 100 Earth masses; Sato et al. 2005). The error bars associated with OGLE-Tr-10b are large enough that its size might agree with theoretical predictions, and recent observations by Holman et al. (2007) indicate that this is actually the case. HD 209458b, on the other hand, is still thought to be anomalously large for its mass, and the reason remains obscure.

A favored generic method for producing large planetary radii is to invoke an internal energy source. One possible source is the dissipation of waves generated by vigorous near-surface flows, which in turn might be driven by the large dayside-to-nightside difference in stellar energy flux (Guillot & Showman 2002). Details of the generation and dissipation mechanisms are unclear, however. One must also ask why, if such a mechanism works efficiently on HD 209458b, it does not do so on any other planet. A different suggestion is that the heat-input method for close-in giant planets requires the presence of another large planet in the system, one able to pump up the eccentricity of the known planet's orbit, which can then cause internal energy dissipation via tidal dissipation. But in the

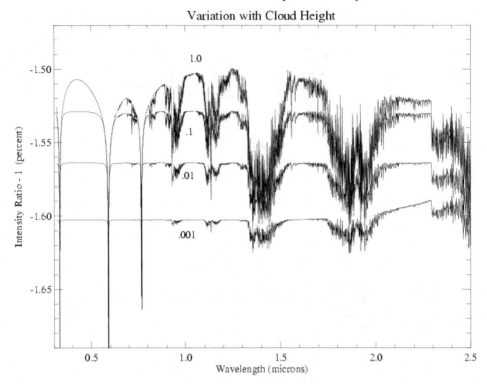

FIGURE 6. Theoretical transit spectra of HD 209458b from Brown (2001), showing the effect of varying the height of an opaque cloud deck. The spectrum features arise mostly from the pressure-broadened lines of alkali metals, and from molecules of $H_2O$, CO, and (to a small extent) $CH_4$.

case of HD 208458b, there is no radial velocity evidence for a second large planet (Mazeh et al. 2000; Naef et al. 2004), and timing measurements of the secondary eclipse (described below) show that either its orbital eccentricity is very small, or that Earth lies very near to its orbital major axis (Deming et al. 2005). The reason for HD 209458b's large radius thus remains mysterious.

### 4.2. *Na and CO*

As described above, atomic or molecular species that have strong absorption lines will cause a transiting planet to appear larger at absorbing wavelengths than at nearby wavelengths, where the opacity is smaller. Many of the strong molecular bands that are expected in the atmospheres of hot Jupiters arise from $H_2O$ and $CH_4$, molecules that are also fairly abundant in the Earth's atmosphere. Thus, telluric obscuration makes these bands difficult to detect from the ground (though the possibility of detecting signals from relatively high-excitation molecular lines should not be discounted—see, e.g., Carr et al. 2004). There are, however, a few species with strong lines that are not fatally contaminated by telluric absorption; these include the alkali metals (especially Na) and the CO molecule.

Sodium is a minor constituent of the Sun, with an abundance of roughly $2 \times 10^{-6}$ by number relative to hydrogen. The opacity of the Na D resonance lines is so large, however, that they are expected to be two of the most prominent lines in the atmosphere of a hot Jupiter. They are also strongly susceptible to pressure broadening, so that at pressures of a bar or so, the pressure-broadened wings of the D lines are the dominant source of opacity for a wide range of wavelengths around the lines themselves. At lower pressures the lines become much narrower. Thus, barring some process that removes neutral sodium atoms from the atmosphere, these lines are expected to be opaque to tangential rays up to heights of perhaps 5000 km (6% of the planetary radius) above the one bar level. In the absence of clouds, the equivalent width of the resulting transmission spectrum features could be quite large, as shown in the top trace in Figure 6.

Motivated by these considerations, Charbonneau et al. (2002) used the wavelength-resolved transit photometry measurements taken with *HST* to measure the difference in transit depth between three increasingly broad regions surrounding the Na D lines and the surrounding continuum. They succeeded in detecting increased absorption in the D lines, but with an increase of transit depth of only $(2.32 \pm 0.57) \times 10^{-4}$ (compared to a total transit depth of 0.0165). The increase in transit depth was measured to be largest in the smallest (1.2 nm wide) wavelength band analyzed, smaller in the intermediate (3.8 nm wide) band, and undetectable in the largest (10 nm wide) band. In the narrowest band, the increase in transit depth was about 0.5 times that expected from a "fiducial" model with a cloud deck topping out at 0.03 bar, indicated by the trace labeled ".1" on Figure 6. This observation remains the only statistically significant ($4\sigma$) measurement of a spectral feature in the transit spectrum of an extrasolar planet. This fact alone should motivate further work on observational transit spectroscopy, though from ground-based telescopes the difficulties involved in such measurements are formidable.

Clouds are not the only mechanism that could result in weak Na D lines. Any process that reduces the abundance of atoms in the lower state of the D line transitions would also reduce the absorption. The magnitude of the Na depletion would have to be quite large, however, by virtue of Eq. (2.2). Indeed, to reduce the equivalent width of the Na feature by the observed factor of two would require depleting the Na abundance by a factor of about 100. Nevertheless, a variety of mechanisms for generating such a depletion have been suggested, including low initial abundance of all heavy elements, and combination of Na atoms to form molecules (such as NaI) that might exist in the solid phase and ultimately rain out of the upper parts of the atmosphere. Perhaps the mechanism that has been best studied quantitatively removes atoms from the D line lower state via non-LTE effects, driven by short-wavelength radiation from the star (Barman et al. 2002). This study suggests that non-LTE corrections to the atomic level populations may be important, particularly high in the atmosphere. The radiative transfer model used is not directly applicable to the transit geometry, however, so some caution is required in interpreting the results.

Among the various molecular features that can be identified in the synthetic spectrum in Figure 6, the most prominent in the wavelength range $1.0 \; \mu\text{m} \leqslant \lambda \leqslant 2.5 \; \mu\text{m}$ are caused by $H_2O$. Because of the already-mentioned time-varying extinction from telluric water vapor, serious efforts to observe any of these molecular bands have so far been made only from space. Grism observations of three transits of HD 209458b searching for $H_2O$ absorption near $\lambda = 1.5 \; \mu\text{m}$ using NICMOS on *HST* have recently been executed. Analysis of these data is underway, but no results are yet reported.

The next most promising molecular signature in the NIR part of the spectrum is the CO bandhead near 2.2 $\mu$m. Although the concentration of CO in the Earth's atmosphere is low enough that it does not cause significant extinction, observations of the CO band are nevertheless severely hampered by overlying lines of $CH_4$ and $H_2O$. Two attempts to detect the CO band in the transit spectrum of HD 209458b have been carried out using the NIRSPEC spectrometer on the Keck II telescope. The first of these was hampered by poor weather, and gave an upper limit on the CO band strength that was larger even than that predicted by cloud-free models (Brown et al. 2002). Although astrophysically uninteresting, these observations suggested that with improved techniques and better weather, a meaningful measurement might be obtained. This goal was achieved by Deming et al. (2005a), who used the same equipment (but with an improved observing strategy) to set an upper limit on the band strength that is smaller by a factor of about six than that predicted by models with clouds up to 0.03 bar. To account for the observed upper limit on band absorption, it is again necessary either to increase the cloud-top height to 3.3 mb or higher, or to invoke some mechanism to drastically reduce the mixing ratio of CO in the planet's upper atmosphere.

## 5. Secondary transits and thermal IR

During the secondary transit of a planet, the planet passes behind the disk of its parent star. During the bulk of such a transit, the planet is completely obscured by the star, resulting in a genuinely flat-bottomed transit light curve, with fairly brief ingress and egress portions during which the planet's disk is in the process of being covered or uncovered. For stars of roughly solar size, planets roughly the size of Jupiter, and orbital periods of a few days, the total transit duration is typically two to three hours, and the duration of the ingress/egress phases is 15 to 20 minutes. The short duration of the ingress/egress phase is important, because it means that instruments to observe the secondary transits need only be photometrically stable for these relatively short times.

In visible light, for which almost all of the light from the planet consists of scattered starlight, the depth of the secondary transit dip is too small to be seen by current techniques. Again taking as typical a planet with radius of $1.2\,R_{\rm Jup}$, orbiting a Sun-like star with a period of four days, one finds the total flux $I_p$ scattered by the planet to be

$$I_p = I_* \frac{1}{2} \frac{R_p^2}{a^2} A \simeq 7.6 \times 10^{-5} A \; , \tag{5.1}$$

where $I_*$ is the flux seen from the star, and $A$ is the geometric albedo of the planet. This estimate depends upon Lambertian scattering from the planet, and one can imagine atmospheric effects ("glories" and related phenomena) that might cause enhanced backscattering when the planet is near its full phase. Nevertheless, this estimate must be roughly correct, and it suggests a signal that is discouragingly small for any ground-based observation. (Note, however, that by utilizing the Doppler shift of the scattered radiation from the planet as it moves in its orbit, it is possible to devise an observational scheme that is differential in wavelength. In this way, detection of such small signals may be possible. So far, attempts by Charbonneau et al. (1999) to apply such a method have resulted in interesting upper limits on the albedo $A$, but no detections.) Space-borne detections may also be possible in the near future (Green et al. 2003), though observations of $\tau$ Boo with the MOST instrument have as yet proved inconclusive (Matthews 2005).

In the thermal infrared, the situation is much better. Recall that for an object with an effective temperature of 1000 K, the peak photon emission of the black body spectrum falls at about 3.7 $\mu$m, so that "thermal" wavelengths can be only a few $\mu$m. In the Rayleigh-Jeans long-wavelength limit, $I_p$ is given by

$$I_p = I_* \frac{R_p^2}{R_*^2} \frac{T_p}{T_*} \; , \tag{5.2}$$

where $T_p$ and $T_*$ are the effective temperatures of the planet and star, respectively. For $T_p = 1100$ K and $T_* = 5800$ K, and $R_p$ the same as above, this gives $I_p \simeq 0.003 I_s$.

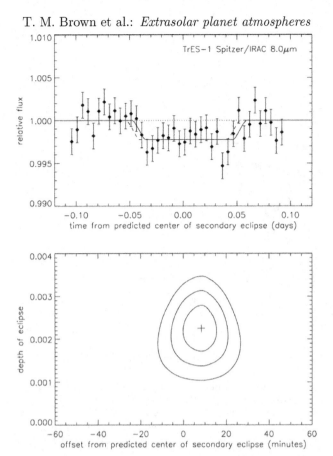

FIGURE 7. IRAC data from Charbonneau et al. (2005), showing (top panel) the 8 $\mu$m flux variation during secondary transit as a function of time, and (bottom panel) the $\chi^2$ surface for fitting the data, as a function of transit center time and transit depth. Contours in the lower panel correspond to confidence levels of 1, 2, and 3 standard deviations.

Though not a huge signal, this ratio is much more promising than the value for reflected light.

Recently such observations of the planets TrES-1 and HD 209458b have been undertaken using the *Spitzer Space Telescope* at wavelengths of 4.5, 8, and 24 $\mu$m (Charbonneau et al. 2005; Deming et al. 2005). The 8 $\mu$m observations of TrES-1 are shown in Figure 7; the top panel shows the observed light curve, while the bottom panel shows the $\chi^2$ surface resulting from a fit to the data, with the transit depth and center time taken as free parameters. Although the fluxes involved are much larger, the 4.5 $\mu$m observations of TrES-1 and the 24 $\mu$m observations of HD 209458b both have similar statistical quality, because the relative transit depths grow larger at wavelengths longer than 8 $\mu$m, and smaller at shorter wavelengths.

The depths of secondary transits are, in effect, measurements of the IR fluxes from the respective planets, averaged over the wavelength bands in question. It is thus possible to compare these fluxes with those predicted by models of the planetary atmospheres (augmented with information about the planets' sizes derived from primary transit data). Such comparisons have been done by Burrows et al. (2005); Fortney et al. (2005), and Seager et al. (2005), with interesting results, as seen in Figure 8, which has been adapted from Burrows et al. (2005).

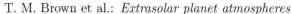

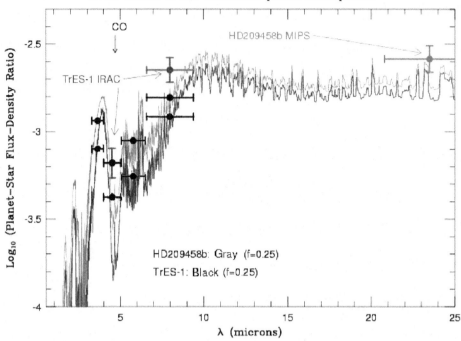

FIGURE 8. Model thermal emission spectra of HD 209458b (upper curve) and TrES-1 (lower curve) as computed by Burrows et al. (2005), overlain with observed fluxes from Charbonneau et al. (2005) and Deming et al. (2005). The observations are shown as points with both vertical and horizontal error bars, the latter representing the wavelength range sampled by the observations. Points with only horizontal bars represent averages of the model fluxes over the indicated wavelength range.

　　Model planetary emission spectra show a marked dip in the flux near 4.6 $\mu$m, due to strong absorption by the CO vibration-rotation band at that wavelength. The relatively low flux observed at 4.5 $\mu$m may, therefore, be interpreted as evidence for the presence of CO in the planetary atmosphere (e.g., Burrows et al. 2005). This result is not inconsistent with the failure of ground-based observations to find evidence for CO, because the 4.6 $\mu$m band is the fundamental vibration-rotation band, which is much stronger than the 2.3 $\mu$m first overtone band. Similarly, the broad dip in the model spectra between 4.5 $\mu$m and 10 $\mu$m results mainly from water vapor. The slightly lower flux at this wavelength compared to 10 $\mu$m is evidence (albeit weak) for water absorption.

　　The observed fluxes at all wavelengths are uniformly greater than those computed in the models by Burrows et al. (2005). These models were computed assuming that the incoming stellar heat flux distributes itself uniformly over the entire surface of the planet, i.e., that heat transport from the planet's day side to its night side is very efficient. Comparison of the observed fluxes with the models by Burrows et al. (2005) therefore suggests that the heat transport to the planet's night side may be quite inefficient, thereby raising the temperature of the planet on the day side. This conclusion appears model dependent, however. Thus, Fortney et al. (2005) find that agreement with their atmospheric models demands efficient heat transport (else the observed fluxes would be larger than they are), while Seager et al. (2005) find an intermediate case, in which the heat transport must be moderately efficient. This disparity among models is a bit discouraging, but it likely is an inevtiable state of affairs, given the untested nature of both the models and the observations.

Finally, the timing of the secondary transits can be used to set limits on the eccentricities of the planetary orbits. If an orbit is elliptical and the Earth does not lie near the projection of its major axis, then the secondary eclipses do not occur midway in time between primary eclipses. This timing difference turns out to be quite large, even for seemingly small eccentricities—as much as 37 minutes for eccentricity of 0.01 and a four-day orbital period. Since the primary transit center times can be estimated with errors of only a few tens of seconds, and the secondary ones to better than 10 minutes, it is possible to set stringent bounds on the possible combination of orbital eccentricity and major axis orientation. For both HD 209458b and TrES-1b, Charbonneau et al. (2005) and Deming et al. (2005) find that the timings are consistent with circular orbits, within the observational errors. If the Earth lay along the minor axes of the orbits, these observations would imply orbital eccentricities of less than about 0.003 for both planets. Since the orientations of the hypothetically elliptical orbits are not known, substantial eccentricities are not formally excluded; they are, however, very unlikely. The small limits on the probable eccentricities strongly suggest that damping of orbital eccentricity does not provide an internal energy source that might cause the inflated radius of HD 209458b.

## 6. Conclusions

A combination of observations and theory is beginning to reveal the nature of the atmospheres of extrasolar giant planets.

Infrared observations of secondary transits hold great promise as diagnostics of the atmospheres, providing information about composition, about dayside temperatures, and, by inference, about heat flow and dynamics. The *Spitzer Space Telescope* provides five bandpasses in which secondary transits of the nearer transiting planets can be detected. Although five-wavelength spectrophotometry of two or three planets is scarcely all that one might desire, it is vastly more information about the planetary atmospheres than has been available hitherto.

High-resolution near-IR spectroscopy of common molecules such as $H_2O$ and CO may yet prove successful, but efforts to date have failed to provide detections of the primary transit spectra of even the most accessible transiting planets. Theory suggests that these spectra contain a lot of information. Adequate observations may however await the launch of space telescopes designed for this purpose; none of these are now on the horizon (although *JWST* holds some promise for at least some kinds of observations). The upper limits on molecular band strengths that have been set using ground-based observations suggest that sources of continuous opacity (clouds, perhaps) are likely important in the atmospheres of hot Jupiters at pressures of a few mb or less.

Visible-light (but space-borne) observations of primary transits have revealed absorption in the Na D lines; the small amplitude of the absorption is consistent with the idea of high clouds, but other explanations are possible as well. The mass-radius relation for extrasolar planets is measured using accurate transit light curves, and it provides constraints on the bulk composition of the planets. Most of the transiting planets agree tolerably well with theoretical expectations for irradiated hydrogen-helium configurations, but the planet orbiting HD 149026 is too small (likely because of a large and dense core), while that orbiting HD 209458 is too large (for reasons that remain mysterious).

Finally, the observations of the transit of Venus described above still await a complete analysis, but a preliminary look shows that the transit spectrum of Venus's atmosphere can easily be seen in the near-IR lines of $CO_2$. These data (and those that will no doubt be obtained at the next Venus transit in 2012), provide both an inspiration and a testing ground for methods that we would like to use on planets of other stars.

We are grateful to the staff of the KIS vacuum tower telescope and to Manuel Collados Vera and the staff of the Instituto de Astrofisica de Canarias for their generous assistance with the observations of the Venus transit. NSO/Kitt Peak FTS data used here were produced by NSF/NOAO.

## REFERENCES

ALONSO, R., ET AL. 2004 *ApJ* **613**, L153.

BARMAN, T. S., HAUSCHILDT, P. H., SCHWEITZER, A., STANCIL, P. C., BARON, E., & ALLARD, F. 2002 *ApJ* **569**, L51.

BODENHEIMER, P., LAUGHLIN, G., & LIN, D. N. C. 2003 *ApJ* **592**, 555.

BROWN, T. M. 2001 *ApJ* **553**, 1006.

BROWN, T. M., CHARBONNEAU, D., GILLILAND, R. L., NOYES, R. W., & BURROWS, A. 2001 *ApJ* **552**, 699.

BROWN, T. M., LIBBRECHT, K. G., & CHARBONNEAU, D. 2002 *PASP* **114**, 826.

BURROWS, A., SUDARSKY, D., & HUBBARD, W. B. 2003 *ApJ* **594**, 545.

BURROWS, A., HUBENY, I., HUBBARD, W. B., SUDARSKY, D., & FORTNEY, J. J. 2004 *ApJ* **610**, L53.

BURROWS, A., HUBENY, I., & SUDARSKY, D. 2005 *ApJ* **625**, L135.

CARR, J. S., TOKUNAGA, A. T., & NAJITA, J. 2004 *ApJ* **603**, 213.

CHABRIER, G., BARMAN, T., BARAFFE, I., ALLARD, F., & HAUSCHILDT, P. H. 2004 *ApJ* **603**, L53.

CHARBONNEAU, D., NOYES, R. W., KORZENNIK, S. G., NISENSON, P., JHA, S., VOGT, S. S., & KIBRICK, R. I. 1999 *ApJ* **522**, L145.

CHARBONNEAU, D., BROWN, T. M., NOYES, R. W., & GILLILAND, R. L. 2002 *ApJ* **568**, 377.

CHARBONNEAU, D., ET AL. 2005 *ApJ* **626**, 523.

DEMING, D., BROWN, T. M., CHARBONNEAU, D., HARRINGTON, J., & RICHARDSON, L. J. 2005 *ApJ* **622**, 1149.

DEMING, D., SEAGER, S., RICHARDSON, L. J., & HARRINGTON, J. 2005 *Nature* **434**, 740.

FISCHER, D. A., & VALENTI, J. 2005 *ApJ* **622**, 1102.

FORTNEY, J. J., MARLEY, M. S., LODDERS, K., SAUMON, D., & FREEDMAN, R. 2005 *ApJ* **627**, L69.

GAUDI, B. S. 2005 *ApJ* **628**, L73.

GREEN, D., MATTHEWS, J., SEAGER, S., & KUSCHNIG, R. 2003 *ApJ* **597**, 590.

GUILLOT, T., & SHOWMAN, A. P. 2002 *A&A* **385**, 156.

HOLMAN, M., ET AL. 2007 *ApJ* **655**, 1103.

KLEY, W., LEE, M. H., MURRAY, N., & PEALE, S. J. 2005 *A&A* **437**, 727.

KONACKI, M., TORRES, G., JHA, S., & SASSELOV, D. D. 2003 *Nature* **421**, 507.

KONACKI, M., TORRES, G., SASSELOV, D. D., & JHA, S. 2005 *ApJ* **624**, 372.

LAUGHLIN, G., BUTLER, R. P., FISCHER, D. A., MARCY, G. W., VOGT, S. S., & WOLF, A. S. 2005 *ApJ* **622**, 1182.

LIVINGSTON, W. & WALLACE, L. 1991 *NSO Technical Report* #91-001.

MARCY, G. W., BUTLER, R. P., FISCHER, D. A., & VOGT, S. S. 2003. In *Scientific Frontiers in Research on Extrasolar Planets* (eds. D. Deming & S. Seager). ASP Conf. Ser. 294, p. 1. ASP.

MATTHEWS, J. 2005 (these proceedings).

MAYOR, M. & QUELOZ, D. 1995 *Nature* **378**, 355.

MAZEH, T., ET AL. 2000 *ApJ* **532**, L55.

MOUTOU, C., PONT, F., BOUCHY, F., & MAYOR, M. 2004 *A&A* **424**, L31.

NAEF, D., MAYOR, M., BEUZIT, J. L., PERRIER, C., QUELOZ, D., SIVAN, J. P., & UDRY, S. 2004 *A&A* **414**, 351.

PONT, F., BOUCHY, F., QUELOZ, D., SANTOS, N. C., MELO, C., MAYOR, M., & UDRY, S. 2004 *A&A* **426**, L15.

SANTOS, N. C., ISRAELIAN, G., MAYOR, M., REBOLO, R., & UDRY, S. 2003 *A&A* **398**, 363.

SATO, B., ET AL. 2005 *ApJ* **633**, 465.

SEAGER, S. & MALLÉN-ORNELAS, G. 2003 *ApJ* **585**, 1038.

SEAGER, S., RICHARDSON, L. J., HANSEN, B. M. S., MENOU, K., CHO, J. Y.-K., & DEMING, D. 2005 *ApJ* **632**, 1122.

SEAGER, S. & SASSELOV, D. D. 2000 *ApJ* **537**, 916.

SOZZETTI, A., ET AL. 2004 *ApJ* **616**, L167.

THOMMES, E. W. & LISSAUER, J. J. 2005. In *Astrophysics of Life* (eds. M. Livio, I. N. Reid, & W. B. Sparks). p. 41. Cambridge University Press.

UDALSKI, A., ET AL. 2002 *Acta Astronomica* **52**, 1.

UDALSKI, A., SZEWCZYK, O., ZEBRUN, K., PIETRZYNSKI, G., SZYMANSKI, M., KUBIAK, M., SOSZYNSKI, I., & WYRZYKOWSKI, L. 2002 *Acta Astronomica* **52**, 317.

WINN, J., ET AL. 2005 *ApJ* **631**, 1215.

# Planetary migration

By **PHILIP J. ARMITAGE**[1,2] AND **W. K. M. RICE**[3]

[1]JILA, 440 UCB, University of Colorado, Boulder, CO 80309, USA

[2]Department of Astrophysical and Planetary Sciences, University of Colorado, Boulder, CO 80309, USA

[3]Institute of Geophysics and Planetary Physics and Department of Earth Sciences, University of California, Riverside, CA 92521, USA

Gravitational torques between a planet and gas in the protoplanetary disk result in orbital migration of the planet, and modification of the disk surface density. Migration via this mechanism is likely to play an important role in the formation and early evolution of planetary systems. For masses comparable to those of observed giant extrasolar planets, the interaction with the disk is strong enough to form a gap, leading to coupled evolution of the planet and disk on a viscous time scale (Type II migration). Both the existence of hot Jupiters and the statistical distribution of observed orbital radii are consistent with an important role for Type II migration in the history of currently observed systems. We discuss the possibility of improving constraints on migration by including information on the host stars' metallicity, and note that migration could also form a population of massive planets at *large* orbital radii that may be indirectly detected via their influence on debris disks. For lower mass planets with $M_p \sim M_\oplus$, surface density perturbations created by the planet are small, and migration in a laminar disk is driven by an intrinsic and apparently robust asymmetry between interior and exterior torques. Analytic and numerical calculations of this Type I migration are in reasonable accord, and predict rapid orbital decay during the final stages of the formation of giant planet cores. The difficulty of reconciling Type I migration with giant planet formation may signal basic errors in our understanding of protoplanetary disks, core accretion, or both. We discuss physical effects that might alter Type I behavior, in particular the possibility that for sufficiently low masses ($M_p \rightarrow 0$), turbulent fluctuations in the gas surface density dominate the torque, leading to random walk migration of very low mass bodies.

## 1. Introduction

The extremely short orbital period of 51 Pegasi (Mayor & Queloz 1995) and the other hot Jupiters pose a problem for planet formation, not only because such systems bear little resemblance to the Solar System, but more fundamentally because the high temperatures expected in the protoplanetary disk at radii $a < 0.1$ AU largely preclude the possibility of in situ formation. Disk models by Bell et al. (1997) show that for typical T Tauri accretion rates of $\dot{M} \sim 10^{-8} \, M_\odot \, \mathrm{yr}^{-1}$ (Gullbring et al. 1998), the midplane temperature interior to 0.1 AU exceeds 1000 K, destroying ices and, for the very closest in planets, even dust. At least the cores of these hot Jupiters must, therefore, have formed elsewhere, and subsequently migrated inward. Migration is also likely to have occurred for the larger population of extrasolar planets that now lie within the snow line in their parent disks (Bodenheimer, Hubickyj, & Lissauer 2000), though this is a more model-dependent statement since both the location of the snow line (Sasselov & Lecar 2000) and its significance for giant planet formation remain uncertain.

Orbital migration of planets involves a loss of angular momentum to either gas or other solid bodies in the system. Three main mechanisms have been proposed, all of which involve purely gravitational interactions (aerodynamic drag, which is central to the orbital evolution of meter-scale rocks, is negligible for planetary masses). The first is gravitational interaction between the planet and the gas in the protoplanetary disk.

This leads to angular momentum exchange between the planet and the gas, and resulting orbital evolution (Goldreich & Tremaine 1980; Lin, Bodenheimer & Richardson 1996). Since gas giant planets, by definition, formed at an epoch when the protoplanetary disk was still gas rich, this type of migration is almost unavoidable. It is the main subject of this article. However, further migration could also occur later on, after the gas disk has been dissipated, as a consequence of the gravitational scattering of either planetesimals (Murray et al. 1998) or other massive planets (Rasio & Ford 1996; Weidenschilling & Marzari 1996; Lin & Ida 1997; Papaloizou & Terquem 2001). Some orbital evolution from planetesimal scattering is inevitable, given that the formation of massive planets is highly likely to leave a significant mass of smaller bodies in orbits close enough to feel perturbations from the newly formed giant. In the Solar System, planetesimal scattering could have allowed substantial outward migration (Thommes, Duncan, & Levison 1999) of Uranus and Neptune—which have a small fraction of the Solar System's angular momentum—while simultaneously raising the eccentricities and inclinations of all the giant planets to values consistent with those observed (Tsiganis et al. 2005). Although this is an attractive theory for the architecture of the outer Solar System, invoking a scaled-up version of this process as the origin of the hot Jupiters is problematic. To drive large-scale migration of the typically rather massive planets seen in extrasolar planetary systems would require a comparable mass of planetesimals interior to the initial orbit of the planet. Such a planetesimal disk would, in turn, imply the prior existence of a rather massive gas disk, which would likely be more effective at causing migration than the planetesimals. Similar reservations apply to models of planet-planet scattering, which is only able to yield a population of planets at small orbital radii if multiple planet formation (with the planets close enough that they are unstable over long periods) is common. That said, the observation that most extrasolar planets have significantly eccentric orbits—which currently defies explanation *except* as an outcome of planet-planet scattering (Ford, Rasio, & Yu 2003)—may mean that at least some scattering-driven migration occurs in the typical system.

Figure 1 illustrates how a planet on a circular orbit interacts with the protoplanetary disk. The planet perturbs the gas as it passes by the planet, with angular momentum transport taking place at the locations of resonances in the disk—radii where a characteristic disk frequency is related to the planet's orbital frequency. For relatively low-mass perturbers, the interaction launches a trailing spiral wave in the gas disk, but is not strong enough to significantly perturb the azimuthally averaged surface-density profile. In this regime, described as Type I migration, angular-momentum transport between the planet and the gas occurs while the planet remains embedded within the protoplanetary disk. The rate of migration is controlled by the sum of the torques arising from the inner and outer Lindblad and corotation resonances, which is generally non-zero (if the sum happened to be close to zero, the planet would act as a source of angular-momentum transport in the disk (Goodman & Rafikov 2001), while remaining in place). For the parameters (sound speed, efficiency of angular-momentum transport) that are believed to be appropriate for protoplanetary disks, Type I migration occurs for planet masses $M_p \lesssim 0.1 \, M_J$, where $M_J$ is the mass of Jupiter (Bate et al. 2003), and is most rapid as this critical mass is approached (e.g., Ward 1997). As a result, it is likely to play a particularly important role in the final assembly of giant planet cores.

At higher masses—$M_p \gtrsim 0.1 \, M_J$—the angular momentum removal/deposition at the planet's inner/outer Lindblad resonances is strong enough to repel gas from an annular region surrounding the planet's orbit, forming a gap in which the surface density is reduced compared to its unperturbed value. For planets of a Jupiter mass and above, the gap is almost entirely evacuated (e.g., the right-hand panel of Figure 1), although

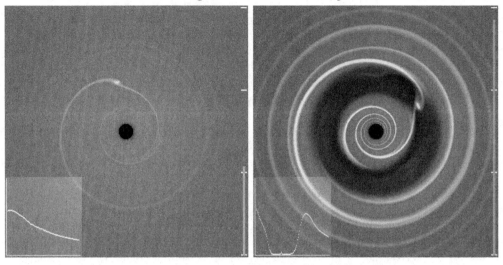

FIGURE 1. An illustration of the interaction between a planet on a fixed circular orbit with a laminar (non-turbulent) protoplanetary disk, computed from a two-dimensional $(r, \phi)$ hydrody-namic simulation with a locally isothermal equation of state and a constant kinematic viscosity. In the left-hand panel showing the regime of Type I migration, a relatively low-mass planet excites a noticeable wave in the disk gas, but does not significantly perturb the azimuthally averaged surface-density profile (shown as the inset graph). In contrast, a 10 $M_J$ planet (right–hand panel) clears an annular gap in the disk, within which the surface density is a small fraction of its unperturbed value. As the disk evolves over a viscous time scale, the planet is predicted to track the motion of the gas (either inward or outward) while remaining within the gap. This is Type II migration.

mass may continue to flow in a stream across the gap to enter the planet's Roche lobe (Artymowicz & Lubow 1996; Lubow, Seibert, & Artymowicz 1999). The location of the inner and outer edges of the gap are set by a balance between angular momentum exchange with the planet (which tends to widen the gap), and internal stresses within the protoplanetary disk ('viscosity,' which tends to close it). Under most circumstances, this balance acts to lock the planet into the long-term viscous evolution of the disk gas. At radii where the disk gas is moving inward, the planet migrates toward the star, all the while remaining within its gap (Lin & Papaloizou 1986). This is Type II migration, which differs from Type I not only in the presence of a gap, but also because the rate depends directly on the efficiency of angular-momentum transport within the protoplanetary disk.

In addition to these well-established migration regimes, qualitatively different behavior may occur at very low masses, and at masses intermediate between the Type I and Type II regimes. At sufficiently low masses (probably of $\sim$10 $M_\oplus$ and below), the persistent Type I torque may be overwhelmed by *random* torques from surface density perturbations in a turbulent protoplanetary disk (Nelson & Papaloizou 2004; Laughlin, Steinacker, & Adams 2004). This process, which is similar to the heating of Galactic stars by transient spiral arms (Carlberg & Sellwood 1985), may lead to rapid random walk migration on top of Type I drift. This could have important consequences for both core accretion (Rice & Armitage 2003) and terrestrial planet formation. Another uncertain regime lies at the transition mass ($M_p \approx 0.1 \ M_J$) between Type I and Type II migration, where a partial gap exists and corotation torques can be highly significant. Masset & Papaloizou (2003) suggested that the corotation torques could drive an instability in the *direction* of migration, which, if confirmed, would be extremely important for massive planet formation. Subsequent higher resolution simulations by D'Angelo, Bate, & Lubow (2005) yield a smaller contribution from the corotation region, and slower migration, but still fail to achieve numerical convergence. The possibility of runaway migration at masses just beyond the Type I threshold remains, therefore, open.

## 2. Type I migration

In the Type I regime, the perturbation induced by the planet in the gas disk remains small, and the net torque has surprisingly little dependence on the microphysics of the protoplanetary disk (e.g., Goldreich & Tremaine 1978; Lin & Papaloizou 1979). In particular, viscosity—normally the most uncertain element of a protoplanetary disk model—enters only indirectly via its influence on the magnitude and radial gradient of the surface density and sound speed. Generically, the net torque scales with the planet mass as $T \propto M_p^2$, so that the migration time scale at a given radius scales as $\tau \propto M_p^{-1}$. Type I migration, therefore, becomes increasingly important as the planet mass increases, and is fastest just prior to gap opening (the onset of which *does* depend on the disk viscosity). Despite its attractive lack of dependence on uncertain disk physics, the actual calculation of the net torque is technically demanding, and substantial improvements have been made only recently. Here, we summarize a few key results—the reader is directed to the original papers (primarily by Artymowicz, Ward, and their collaborators) for full details of the calculations.

### 2.1. Analytic calculations

The simplest calculation of the Type I torque (Goldreich & Tremaine 1978, 1979, 1980) neglects significant pressure effects in the disk close to the planet, and is therefore valid for low $m$ resonances. In this approximation, Lindblad resonances occur at radii in the disk where the epicyclic frequency $\kappa$ is an integral multiple $m$ of the angular velocity in a frame rotating with the planet at angular velocity $\Omega_p$. For a Keplerian disk, this condition,

$$D(r) \equiv \kappa^2 - m^2 (\Omega - \Omega_p)^2 = 0 \ ,$$

can be simplified using the fact that $\kappa = \Omega$. The resonances lie at radii,

$$r_L = \left(1 \pm \frac{1}{m}\right)^{2/3} r_p \ ,$$

where $r_p$ is the planet's orbital radius. The lowest order resonances lie at $r = 1.587 r_p$ and $r = 0.630 r_p$, but an increasingly dense array of high $m$ resonances lie closer to the planet. Resonances at $r < r_p$ add angular momentum to the planet, while those at $r > r_p$ remove angular momentum. The torque at each resonance $T_m$ can be evaluated in terms of a *forcing function* $\Psi_m$ as,

$$T_m = -\pi^2 m \Sigma \frac{\Psi_m^2}{r \mathrm{d}D/\mathrm{d}r} \ ,$$

where $\Sigma$ is the gas surface density. Explicit expressions for $\Psi_m$ are given by Goldreich & Tremaine (1979). The net torque is obtained by evaluating the torque at each resonance, and then summing over all $m$.

For Type I migration the behavior of gas close to the planet, where $\Psi_m$ is largest, is critical. Accurately treating this regime requires elaborations of the basic Goldreich & Tremaine (1979) approach. For a razor-thin two-dimensional disk model—the effects of radial pressure and density gradients—the calculation is described in Ward (1997), and references therein (especially Ward 1988; Artymowicz 1993a, 199b; Korycansky & Pollack 1993). These papers include the shifts in the location of Lindblad resonances due to both radial and azimuthal pressure gradients, which become significant effects at high $m$. For Lindblad resonances, the result is that the dominant torque arises from wavenumbers $m \simeq r_p/h$, where $h$, the vertical disk scale height, is given in terms of the local sound speed $c_s$ by $h = c_s/\Omega_p$. The fractional net torque $2|T_{inner} + T_{outer}|/(|T_{inner}| + |T_{outer}|)$ can be as large as 50% (Ward 1997), with the outer resonances dominating and driving rapid inward migration. Moreover, the small shifts in the locations of resonances that occur in disks with different radial surface density profiles conspire so that the net torque is only weakly dependent on the surface density profile. This means that the predicted rapid inward migration occurs for essentially any disk model in which the sound speed decreases with increasing radius (Ward 1997). Corotation torques—which vanish in the oft-considered disk models with $\Sigma \propto r^{-3/2}$—can alter the magnitude of the torque but are not sufficient to reverse the sign (Korycansky & Pollack 1993; Ward 1997).

The observation that the dominant contribution to the total torque comes from gas that is only $\Delta r \simeq h$ away from the planet immediately implies that a two-dimensional representation of the disk is inadequate, even for protoplanetary disks which are geometrically thin by the usual definition ($h/r < 0.1$, so that pressure gradients, which scale as order $(h/r)^2$, are sub-percent level effects). Several new physical effects come into play in a three-dimensional disk:

1. The perturbing potential has to be averaged over the vertical thickness of the disk, effectively reducing its strength for high $m$ resonances (Miyoshi et al. 1999).

2. The variation of the scale height with radius decouples the radial profile of the midplane density from that of the surface density.

3. Wave propagation in three-dimensional disks is fundamentally different from that in two dimensions, if the vertical structure of the disk departs from isothermality (Lubow & Ogilvie 1998; Bate et al. 2002).

Tanaka, Takeuchi, & Ward (2002) have computed the interaction between a planet and a three-dimensional isothermal disk, including the first two of the above effects. Both Lindblad and corotation torques were evaluated. They find that the net torque is reduced by a factor of 2–3 as compared to a corresponding two-dimensional model, but that migration remains inward and is typically rapid. Specifically, defining a local migration time scale via,

$$\tau = \frac{r_p}{-\dot{r}_p} \quad ,$$

Tanaka, Takeuchi, & Ward (2002) find that for a disk in which $\Sigma \propto r^{-\beta}$ ,

$$\tau = (2.7 + 1.1\beta)^{-1} \frac{M_*}{M_p} \frac{M_*}{\Sigma r_p^2} \left( \frac{c_s}{r_p \Omega_p} \right)^2 \Omega_p^{-1} \quad .$$

As expected, the time scale is inversely proportional to the planet mass and the local surface density. Since the bracket is $\sim (h/r_p)^2$, the time scale also decreases quite rapidly for thinner disks, reflecting the fact that the peak torque arises from closer to the planet as the sound speed drops.

Although still limited by the assumption of isothermality, the above expression represents the current 'standard' estimate of the Type I migration rate. Although slower than two-dimensional estimates, it is still rapid enough to pose a potential problem for planet formation via core accretion. There is, therefore, interest in studying additional physical effects that might reduce the rate further. The influence of disk turbulence is discussed more fully in the next section; here we list some of the other effects that might play a role:

1. **Realistic disk structure models.** The run of density and temperature in the midplane of the protoplanetary disk is not a smooth power-law due to sharp changes in opacity and, potentially, the efficiency of angular momentum transport (Gammie 1996). Menou & Goodman (2004) have calculated Type I rates in Shakura-Sunyaev type disk models, and find that even using the standard Lindblad torque formula there exist regions of the disk where the migration rate is locally slow. Such zones could be preferred sites of planet formation.

2. **Thermal effects.** Jang-Condell & Sasselov (2005) find that the dominant non-axisymmetric thermal effect arises from changes to the stellar illumination of the disk surface in the vicinity of the planet. This effect is most important at large radii, and can increase the migration time scale by up to a factor of two at distances of a few AU.

3. **Wave reflection.** The standard analysis assumes that waves propagate away from the planet, and are dissipated before they reach boundaries or discontinuities in disk properties that might reflect them back toward the planet. Tanaka, Takeuchi, & Ward (2002) observe that reflection off boundaries has the potential to substantially reduce the migration rate. We note, however, that relaxation of vertical isothermality will probably lead to wave dissipation in the disk atmosphere within a limited radial distance (Lubow & Ogilvie 1998), and thereby reduce the possibility for reflection.

4. **Accretion.** Growth of a planet during Type I migration is accompanied by a non-resonant torque, which has been evaluated by Nelson & Benz (2003a). If mass is able to accrete freely onto the planet, Bate et al. (2003) find from three-dimensional simulations that $\dot{M}_p \propto M_p$ for $M_p \lesssim 10\ M_\odot$, with a mass-doubling time that is extremely short (less than $10^3$ yr). In reality, it seems likely the planet will be unable to accept mass at such a rapid rate, so the mass-accretion rate and resulting torque will then depend on the planet structure.

5. **Magnetic fields.** The dominant field component in magnetohydrodynamic disk turbulence initiated by the magnetorotational instability is toroidal (Balbus & Hawley 1998). Terquem (2003) finds that gradients in plausible toroidal magnetic fields can significantly alter the Type I rate, and in some circumstances (when the field decreases rapidly with $r$) stop migration. More generally, a patchy and variable toroidal field might lead to rapid variations in the migration rate. Whether this, or density fluctuations induced by turbulence, is the primary influence of disk fields is unclear.

6. **Multiple planets.** The interaction between multiple planets has not been studied in detail. Thommes (2005) notes that low-mass planets, which would ordinarily suffer rapid Type I migration, can become captured into resonance with more massive bodies that are themselves stabilized against rapid decay by a gap. This may be important for understanding multiple planet formation (and we have already noted that there is circumstantial evidence that multiple massive-planet formation may be common), though it does not explain how the *first* planet to form can avoid rapid Type I inspiral.

7. **Disk Eccentricity.** Type I migration in an axisymmetric disk is likely to damp planetary eccentricity. However, it remains possible that the protoplanetary disk itself might be spontaneously unstable to development of eccentricity (Ogilvie 2001). Papaloizou (2002) has shown that Type I migration can be qualitatively altered, and even reversed, if the background flow is eccentric.

## 2.2. *Numerical simulations*

Hydrodynamic simulations of the Type I regime within a shearing-sheet geometry have been presented by Miyoshi et al. (1999), and in cylindrical geometry by D'Angelo, Henning, & Kley (2002), D'Angelo, Kley, & Henning (2003), and Nelson & Benz (2003b), with the latter paper focusing on the transition between Type I and Type II behavior. The most comprehensive work to date is probably that of Bate et al. (2003), who simulated in three dimensions the interaction with the disk of planets in the mass range $1 \, M_\oplus \leqslant M_p \leqslant 1 \, M_J$. The disk model had $h/r = 0.05$, a Shakura-Sunyaev (1973) $\alpha$ parameter $\alpha = 4 \times 10^{-3}$, and a fixed locally isothermal equation of state (i.e., $c_s = c_s(r)$ only). This setup is closely comparable to that assumed in the calculations of Tanaka, Takeuchi, & Ward (2002), and very good agreement was obtained between the simulation results and the analytic migration time scale. Based on this, it seems reasonable to conclude that *within the known limitations imposed by the restricted range of included physics*, current calculations of the Type I rate are technically reliable. Given this, it is interesting to explore the consequences of rapid Type I migration for planet formation itself.

## 2.3. *Consequences for planet formation*

The inverse scaling of the Type I migration time scale with planet mass means that the most dramatic effects for planet formation occur during the growth of giant plant cores via core accretion (Mizuno 1980). In the baseline calculation of Pollack et al. (1996), which does not incorporate migration, runaway accretion of Jupiter's envelope is catalyzed by the slow formation of an $\approx 20 \, M_\oplus$ core over a period of almost 10 Myr. This is in conflict with models by Guillot (2004), which show that although Saturn has a core of around 15 $M_\oplus$, Jupiter's core is observationally limited to at most $\approx 10 \, M_\odot$, leading to discussion at this meeting of several ways to reduce the theoretically predicted core mass. Irrespective of the uncertainties, however, it seems inevitable that planets forming via core accretion pass through a relatively slow stage in which the growing planet has a mass of 5–10 $M_\oplus$. This stage is vulnerable to Type I drift.

Figure 2 shows the migration time scale for a 10 $M_\oplus$ planet within gas disks with surface-density profiles of $\Sigma \propto r^{-1}$ (very roughly that suggested by theoretical models, e.g., Bell et al. 1997) and $\Sigma \propto r^{-3/2}$ (the minimum mass Solar Nebula profile of Weidenschilling 1977). We consider disks with integrated gas masses (out to 30 AU) of 0.01 $M_\odot$ and 1 $M_J$. The latter evidently represents the absolute minimum gas mass required to build Jupiter or a typical extrasolar giant planet. The torque formula of Tanaka, Takeuchi, & Ward (2002) is used to calculate the migration time scale $\tau$. As is obvious from the figure, migration from 5 AU on a time scale of 1 Myr—significantly less than either the typical disk lifetime (Haisch, Lada & Lada 2001) or the duration of the slow phase of core accretion—is inevitable for a core of mass 10 $M_\oplus$, even if there is only a trace of gas remaining at the time when the envelope is accreted. For more reasonable gas masses, the typical migration time scale at radii of a few AU is of the order of $10^5$ yr. Another representation of this is to note that in the giant planet forming region, we predict significant migration ($\tau = 10$ Myr) for masses $M_p \gtrsim 0.1 \, M_\oplus$, and rapid migration ($\tau = 1$ Myr) for $M_p \gtrsim 1 \, M_\oplus$. We can also plot, for the same disk model, an estimate of the isolation mass (Lissauer 1993; using the gas to planetesimal surface density scaling of Ida & Lin 2004a). The isolation mass represents the mass a growing planet can attain by consuming only those planetesimals within its feeding zone—as such, it is reached relatively rapidly in planet formation models. Outside the snow line, the migration time scale for a planet (or growing core) at the isolation mass is typically a few Myr, reinforcing the conclusion that migration is inevitable in the early stage of giant planet formation. By contrast, in the terrestrial planet region, interior to the snow line, planets need to grow significantly beyond isolation before rapid migration ensues. It is therefore possible for the early stages of terrestrial planet formation to occur in the

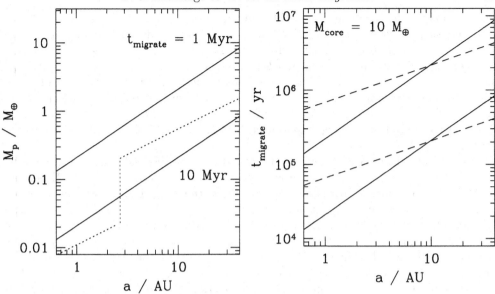

FIGURE 2. Left hand panel: planet mass for which the Type I migration time scale at different radii equals 1 Myr (upper solid curve) and 10 Myr (lower curve). The gas disk is assumed to have a surface density profile $\Sigma \propto r^{-3/2}$, and a mass within 30 AU of 0.01 $M_\odot$. The dotted line shows an estimate of the isolation mass in the same disk model, assuming Solar metallicity and a snow line at 2.7 AU. Right hand panel: the migration time scale for a 10 $M_\oplus$ core in protoplanetary disks with surface density profiles of $\Sigma \propto r^{-3/2}$ (solid lines) and $\Sigma \propto r^{-1}$ (dashed lines). For each model, the lower curve shows the time scale for a disk with a gas mass of 0.01 $M_\odot$ within 30 AU, while the upper curve shows results for an absolute minimum mass gas disk with only 1 $M_J$ within 30 AU.

presence of gas with only limited Type I migration, while the final assembly of terrestrial planets happens subsequently in a gas-poor environment.

Does Type I migration help or hinder the growth of giant planets via core accretion? This question remains open, despite a history of investigations stretching back at least as far as papers by Hourigan & Ward (1984) and Ward (1989). Two competing effects are at work:

1. A migrating core continually moves into planetesimal-rich regions of the disk that have not been depleted by the core's prior growth. This depletion is, in part, responsible for the slow growth of Jupiter in static core calculations. Calculations suggest that a rapidly migrating core can capture of the order of 10% of the planetesimals it encounters (Tanaka & Ida 1999), with the collision fraction increasing with migration velocity. Slow migration velocities allow for inward shepherding of the planetesimals rather than capture, and a low accretion rate (Ward & Hahn 1995; Tanaka & Ida 1999).

2. A migrating core must reach the critical core mass before it is lost to the star, on a time scale that, as we noted above, can be an order of magnitude or more smaller than the gas disk lifetime. Unfortunately, a high accretion rate of planetesimals increases the critical core mass needed before runaway gas accretion starts and, at a fixed accretion rate, the critical mass also increases as the core moves inward (Papaloizou & Terquem 1999). Migration, therefore, favors a high rate of planetesimal accretion, but often hinders attaining the critical core mass needed for envelope accretion.

Calculations of giant planet formation including steady core migration have been presented by several groups, including Papaloizou & Terquem (1999), Papaloizou & Larwood

(2000), Alibert, Mordasini, & Benz (2004) and Alibert et al. (2005). The results suggest that, for a single growing core, Type I migration at the standard rate of Tanaka, Takeuchi, & Ward (2002) is simply too fast to allow giant planet formation to occur across a reasonable range of radii in the protoplanetary disk. Most cores are lost to the star or, if they manage to accrete envelopes at all, do so at such small radii that their ultimate survival is doubtful. More leisurely migration, on the other hand, at a rate suppressed from the Tanaka, Takeuchi, & Ward (2002) value by a factor of 10 to 100, *helps* core accretion by mitigating the depletion effect that acts as a bottleneck for a static core (Alibert et al. 2005).

This difficulty in reconciling our best estimates of the Type I migration rate with core accretion signals that something is probably wrong with one or both of these theories. Three possibilities suggest themselves. First, the Type I migration rate may be a substantial overestimate, by an order of magnitude or more. If so, there is no need for substantial changes to core-accretion theory or to protoplanetary-disk models. We have already enumerated a list of candidate physical reasons for why the Type I rate may be wrong, though achieving a sufficiently large suppression does not seem to be straightforward. Second, angular-momentum transport may be strongly suppressed in the giant-planet formation region by the low ionization fraction, which suppresses MHD instabilities that rely on coupling between the gas and the magnetic field (Gammie 1996; Sano et al. 2000). An almost inviscid disk could lower the threshold for gap opening sufficiently far that the slow stage of core accretion occurred in a gas-poor environment (elements of such a model have been explored by Rafikov 2002; Matsumara & Pudritz 2005). Finally, and perhaps most attractively, it may be possible to find a variant of core accretion that is compatible with undiluted Type I migration. For a single core, we have studied simple models in which random walk migration leads to large fluctuations in the planetesimal accretion rate and an early onset of criticality (Rice & Armitage 2003). In the more realistic situation where multiple cores are present in the disk, it is possible that the early loss of the first cores to form (at small radii just outside the snow line) could evacuate the inner disk of planetesimals, allowing subsequent cores to reach their critical mass and accrete envelopes as they migrate inward. Further work is needed to explore such scenarios quantitatively.

## 3. Stochastic migration in turbulent disks

To a first approximation, the efficiency of angular-momentum transport (unless it is very low) has little impact on the predicted Type I migration rate. This assumes, however, that the disk is laminar. More realistically, angular-momentum transport itself derives from turbulence, which is accompanied by a spatially and temporally varying pattern of density fluctuations in the protoplanetary disk. These fluctuations will exert *random* torques on planets of any mass embedded within the disk, in much the same way as transient spiral features in the Galactic disk act to increase the velocity dispersion of stellar populations (Carlberg & Sellwood 1985). If we assume that the random torques are uncorrelated with the presence of a planet, then the random torques' linear scaling with planet mass will dominate over the usual Type I torque (scaling as $M_p^2$) for sufficiently low masses. The turbulence will then act to increase the velocity dispersion of collisionless bodies, or, in the presence of damping, to drive a random walk in the semi-major axis of low mass planets.

To go beyond such generalities, and in particular to estimate the crossover mass between stochastic and Type I migration, we need to specify the source of turbulence in the protoplanetary disk. MHD disk turbulence (see Figure 3 for an illustration) driven

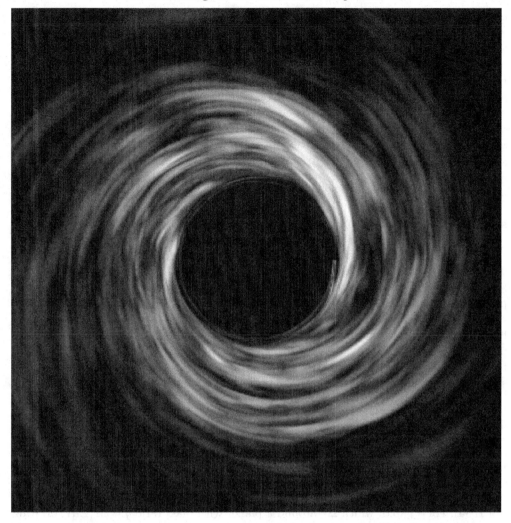

FIGURE 3. Structure in a turbulent disk accretion flow, from a global cylindrical MHD simulation (after Armitage 1998). The flow is visualized here using the square root of the vertically-averaged Maxwell stress $\langle B_r B_\phi \rangle$ as a tracer. In magnetically active disks, transient density and magnetic field fluctuations are present across a wide range of spatial scales.

by the magnetorotational instability (Balbus & Hawley 1998) provides a well-understood source of outward angular-momentum transport in sufficiently well-ionized disks, and has been used as a model system for studying stochastic migration by Nelson & Papaloizou (2004) and by Laughlin, Steinacker, & Adams (2004). Density fluctuations in MHD disk turbulence have a typical coherence time of the order of the orbital period, and, as a consequence, are able to exchange angular momentum with an embedded planet across a range of disk radii (not only at narrow resonances). The study by Nelson & Papaloizou (2004) was based on both global ideal MHD disk simulations, with an aspect ratio of $h/r = 0.07$, and local shearing box calculations. The global runs realized an effective Shakura-Sunyaev $\alpha = 7 \times 10^{-3}$, which, if replicated in a real disk, would be consistent with observational measures of T Tauri disk evolution (Hartmann et al. 1998). For all masses considered in the range $3 \, M_\oplus \leqslant M_p \leqslant 30 \, M_\oplus$, the *instantaneous* torque on the planet from the MHD turbulent disk exhibited large fluctuations in both magnitude and sign. Averaging over $\approx 20$ orbital periods, the mean torque showed signs of converging to the Type I rate, although the rate of convergence was slow, especially for the lowest-mass planets in the global runs. These properties are generally in accord with other studies of the variability of MHD disk turbulence (Hawley 2001; Winters, Balbus & Hawley 2003a). Very roughly, the Nelson & Papaloizou (2004) simulations suggest that up to $M_p \sim 10\ M_\oplus$ the random walk component dominates steady Type I drift over time scales that substantially exceed the orbital period.

We caution that existing studies of the stochastic migration regime are unrealistic. Ideal MHD is not a good approximation for the protoplanetary disk at those radii where planet formation occurs, and there may be dead zones in which MHD turbulence and angular-momentum transport is highly suppressed (Gammie 1996; Sano et al. 2000; Fromang, Terquem, & Balbus 2002; Salmeron & Wardle 2005). We also note that a significant random migration component, if it indeed adds to rather than supplanting steady Type I migration, does nothing (on average) to help the survival prospects of low-mass planets. Nevertheless, if turbulent fluctuations (whatever their origin) do occur in the disk, the resulting random walk migration could be important for planet formation. In the terrestrial-planet region, stochastic migration might deplete low-mass planetary embryos that would be relatively immune to ordinary Type I migration, while simultaneously promoting radial mixing and collision of planetesimals. For giant-planet formation, a significant random component to core migration would have the effect of creating large fluctuations in the planetesimal accretion rate, while also potentially acting to diffuse the planetesimal surface density. Large fluctuations in the planetesimal accretion rate favor the early onset of rapid gas accretion, and allow for the final core mass to be substantially smaller than would be expected in the case of a static core (Rice & Armitage 2003).

## 4. Type II migration

### 4.1. *Conditions for the onset of Type II migration*

In a viscous disk, the threshold between Type I and Type II migration can be derived by equating the time scale for Type I torques to open a gap (in the absence of viscosity) with the time scale for viscous diffusion to fill it in (Goldreich & Tremaine 1980; Papaloizou & Lin 1984). For a gap of width $\Delta r$ around a planet with mass ratio $q = M_p/M_*$, orbiting at distance $r_p$, Type I torques can open the gap on a timescale (Takeuchi, Miyama, & Lin 1996)

$$\tau_{\text{open}} \sim \frac{1}{m^2 q^2} \left(\frac{\Delta r}{r_p}\right)^2 \Omega_p^{-1} \ .$$

Viscous diffusion will close the gap on a time scale,

$$\tau_{\text{close}} \sim \frac{\Delta r^2}{\nu} \ ,$$

where $\nu$, the kinematic viscosity, is usually written as $\nu = \alpha c_s^2/\Omega_p$ (Shakura & Sunyaev 1973). Equating $\tau_{\text{open}}$ and $\tau_{\text{close}}$, and noting that waves with $m \approx r_p/h$ dominate the Type I torque, the condition for gap opening becomes,

$$q \gtrsim \left(\frac{h}{r}\right)_p^2 \alpha^{1/2} \ .$$

For $h/r \simeq 0.05$ and $\alpha = 10^{-2}$, the transition (which simulations show is not very sharp) occurs at $q_{\text{crit}} \sim 2.5 \times 10^{-4}$, i.e., close to a Saturn mass for a Solar-mass star. Since the most rapid Type I migration occurs when $q \approx q_{\text{crit}}$, we can also estimate a *minimum* migration time scale by combining the above expression with the time scale formula of Tanaka, Takeuchi, & Ward (2002). This yields,

$$\tau_{\min} \sim (2.7 + 1.1\beta)^{-1} \frac{M_*}{\Sigma r_p^2} \alpha^{-1/2} \Omega_p^{-1} \quad,$$

and is almost independent of disk properties other than the local mass.

### 4.2. The rate of Type II migration

Once a planet has becomes massive enough to open a gap, orbital evolution is predicted to occur on the same local time scale as the protoplanetary disk. The radial velocity of gas in the disk is,

$$v_r = -\frac{\dot{M}}{2\pi r \Sigma} \quad,$$

which for a steady disk away from the boundaries can be written as,

$$v_r = -\frac{3}{2}\frac{\nu}{r} \quad.$$

If the planet enforces a rigid tidal barrier at the outer edge of the gap, then evolution of the disk will force the orbit to shrink at a rate $\dot{r}_p \simeq v_r$, provided that the local disk mass exceeds the planet mass, i.e. $\pi r_p^2 \Sigma \gtrsim M_p$. This implies a nominal Type II migration time scale, valid for *disk dominated migration* only,

$$\tau_0 = \frac{2}{3\alpha} \left(\frac{h}{r}\right)_p^{-2} \Omega_P^{-1} \quad.$$

For $h/r = 0.05$ and $\alpha = 10^{-2}$, the migration time scale at 5 AU is of the order of 0.5 Myr.

In practice, the assumption that the local disk mass exceeds that of the planet often fails. For example, a $\beta = 1$ disk with a mass of 0.01 $M_\odot$ within 30 AU has a surface density profile,

$$\Sigma = 470 \left(\frac{r}{1\,\text{AU}}\right)^{-1} \text{gcm}^{-2} \quad.$$

The condition that $\pi r_p^2 \Sigma = M_p$ gives an estimate of the radius within which disk domination ceases of,

$$r = 6 \left(\frac{M_p}{M_J}\right) \text{AU} \quad.$$

Interior to this radius, the planet acts as a slowly moving barrier which impedes the inflow of disk gas. Gas piles up behind the barrier—increasing the torque—but this process does not continue without limit because the interaction also deposits angular momentum into the disk, causing it to expand (Pringle 1991). The end result is to slow migration compared to the nominal rate quoted above. For a disk in which the surface density can be written as a power-law in accretion rate and radius,

$$\Sigma \propto \dot{M}^a r^b \quad,$$

Syer & Clarke (1995) define a measure of the degree of disk dominance,

$$B \equiv \frac{4\pi r_p^2 \Sigma}{M_p} \quad.$$

Then for $B < 1$ (the planet dominated case appropriate to small radii) the actual Type II migration rate is (Syer & Clarke 1995),

$$\tau_{II} = \tau_0 B^{-a/(1+a)} \quad.$$

Note that with this definition of $B$, disk dominance extends inward a factor of a few further than would be predicted based on the simple estimate given above.

Evaluating how the surface density depends upon the accretion rate—and thereby determining the $a$ which enters into the suppression term—requires a full model for the protoplanetary disk (not just knowledge of the slope of the steady-state surface density profile). For the disk models of Bell et al. (1997), we find that $a \simeq 0.5$ at 1 AU for $\dot{M} \sim 10^{-8} \, M_\odot \, \mathrm{yr}^{-1}$. At this radius, the model with $\alpha = 10^{-2}$ has a surface density of about 200 gcm$^{-2}$. For a Jupiter-mass planet we then have $B = 0.3$ and $\tau_{II} = 1.5\tau_0$. This is a modest suppression of the Type II rate, but the effect becomes larger at smaller radii (or for more massive planets). It slows the inward drift of massive planets, and allows a greater chance for them to be stranded at sub-AU scales as the gas disk is dissipated.

These estimates of the Type II migration velocity assume that once a gap has been opened, the planet maintains an impermeable tidal barrier to gas inflow. In fact, simulations show that planets are able to accrete gas via tidal streams that bridge the gap (Lubow, Seibert & Artymowicz 1999). The effect is particularly pronounced for planets only just massive enough to open a gap in the first place. If the accreted gas does not have the same specific angular momentum as the planet, this constitutes an additional accretion torque in addition to the resonant torque. It is likely to reduce the Type II migration rate further.

### 4.3. *Simulations*

Simulations of gap opening and Type II migration have been presented by a large number of authors, with recent examples including Bryden et al. (1999), Lubow, Seibert, & Artymowicz (1999), Nelson et al. (2000), Kley, D'Angelo, & Henning (2001), Papaloizou, Nelson, & Masset (2001), D'Angelo, Henning, & Kley (2002), D'Angelo, Kley, & Henning (2003), D'Angelo, Henning, & Kley (2003), Bate et al. (2003), Schäfer et al. (2004) and Lufkin et al. (2004). These authors all assumed, for simplicity, that angular momentum transport in the protoplanetary disk could be represented using a microscopic viscosity. Only a few recent simulations, by Winters, Balbus, & Hawley (2003b), Nelson & Papaloizou (2004) and Papaloizou, Nelson, & Snellgrove (2004), have directly simulated the interaction of a planet with a turbulent disk. For planet masses significantly above the gap opening threshold, simulations support the general scenario outlined above, while also finding:

1. **Significant mass accretion** across the gap. For planet masses close to the gap opening threshold, accretion is surprisingly efficient, with tidal streams delivering gas at a rate that is comparable to the *disk* accretion rate in the absence of a planet (Lubow, Seibert, & Artymowicz 1999). It is also observed that accretion cuts off rapidly as the planet mass increases toward 10 $M_J$, giving additional physical motivation to the standard dividing line between massive planets and brown dwarfs.

2. **Damping of eccentricity**. Goldreich & Tremaine (1980) noted that the evolution of the eccentricity of a planet embedded within a disk depends upon a balance between Lindblad torques, which tend to excite eccentricity, and corotation torques, which damp it. Recent analytic work (Ogilvie & Lubow 2003; Goldreich & Sari 2003; and references therein) has emphasized that if the corotation resonances are even partially saturated, the overall balance tips to eccentricity excitation. To date, however, numerical simulations (e.g., those by Papaloizou, Nelson & Masset 2001) exhibit damping for planetary mass bodies, while eccentricity growth is obtained for masses $M_p \gtrsim 20 \, M_J$—in the brown dwarf regime. In a recent numerical study, however, Masset & Ogilvie (2004) present evidence that the resolution attained by Papaloizou, Nelson, & Masset (2001) was probably

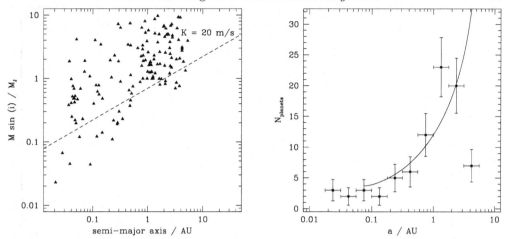

FIGURE 4. Left hand panel: the distribution of extrasolar planets discovered via radial velocity surveys in the $a$–$M_p \sin(i)$ plane. The dashed diagonal traces a line of equal ease of detectability—planets on circular orbits lying along lines parallel to this cause the same amplitude of stellar radial velocity variations. Right hand panel: the number of detected planets with $M_p \sin(i) > 1\ M_J$ is plotted as a function of semi-major axis. The solid curve shows the predicted distribution from a pure migration model by Armitage et al. (2002). The theoretical curve is unaltered from the 2002 version except for an arbitrary normalization.

inadequate to determine the sign of eccentricity evolution for Jupiter mass planets, which remains uncertain. Further numerical work is needed.

### 4.4. *Comparison with statistics of extrasolar planetary systems*

The estimated time scale for migration of a giant planet from 5 AU to the hot Jupiter region is of the order of a Myr. This time scale is short enough—compared to the lifetime of typical protoplanetary disks—to make migration a plausible origin for hot Jupiters, while not being so short as to make large-scale migration inevitable (the latter would raise obvious concerns as to why there is no evidence for substantial migration of Jupiter itself). Having passed this initial test, it is then of interest to try and compare *quantitative* predictions of migration, most obviously the expected distributions of planets in mass and orbital radius, with observations. Models that attempt this include those by Trilling et al. (1998), Armitage et al. (2002), Trilling, Lunine, & Benz (2002), Ida & Lin (2004a) and Ida & Lin (2004b). Accurate knowledge of the biases and selection function of the radial velocity surveys is essential if such exercises are to be meaningful, making analyses such as those of Cumming, Marcy, & Butler (1999) and Marcy et al. (2005) extremely valuable.

Figure 4 shows how the distribution of known extrasolar planets with semi-major axis compares to the pure migration model of Armitage et al. (2002). In this model, we assumed that giant planets form and gain most of their mass at an orbital radius (specifically 5 AU) beyond where most extrasolar planets are currently being detected. Once formed, planets migrate inward via Type II migration and are either (a) swallowed by the star, or (b) left stranded at some intermediate radius by the dispersal of the protoplanetary disk. We assumed that disk dispersal happens as a consequence of photoevaporation (Johnstone, Hollenbach, & Bally 1998), and that the probability of planet formation per unit time is constant over the (relatively short) window during which a massive planet can form at 5 AU and survive without being consumed by the star. Although clearly oversimplified, it is interesting that this model continues to reproduce the orbital dis-

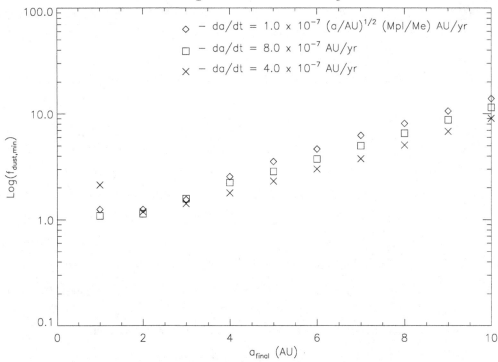

FIGURE 5. The minimum metallicity required to form a gas giant planet as a function of orbital radius, from simplified core accretion calculations by Rice & Armitage (2005). Host metallicity, expressed by the parameter $f_{\rm dust}$, is proportional to $10^{[{\rm Fe}/{\rm H}]}$. The models assume that dispersion in disk metallicity dominates over dispersion in either disk gas mass or disk lifetime in controlling the probability of planet formation. Type I migration of the core prior to accretion of the envelope is included, using several different prescriptions indicated by the different symbols, though none of the Type I rates are as large as the baseline rate of Tanaka, Takeuchi, & Ward (2002).

tribution of known planets out to radii of a few AU, once selection effects have been taken into account (here, by simply ignoring low-mass planets that are detectable only at small orbital radii). Moreover, it predicts that substantial *outward* migration ought to occur in disks where strong mass loss prompts an outwardly directed radial velocity in the giant planet forming region (Veras & Armitage 2004). Planets at these large radii are potentially detectable today via their effect on debris disks (Kuchner & Holman 2003).

Additional clues to the role of migration in forming the observed population of extra-solar planets may be possible by combining large planet samples with knowledge of the host stars' metallicity. It is now clear that the frequency of detected planets increases rapidly with host metallicity, and that the measured metallicity reflects primarily the primordial composition of the star-forming gas rather than subsequent pollution of the convective envelope (Santos, Israelian, & Mayor 2004; Fischer & Valenti 2005; and references therein). The existence of such a correlation is not in itself surprising, given that the time scale for core accretion decreases quickly with increasing surface density of planetesimals. However, the sharpness of the rise in planet frequency over a fairly narrow range of stellar [Fe/H] *is* striking, since it suggests that metallicity, rather than variations in initial gas disk mass or gas disk lifetime, may well be the single most important parameter determining the probability of giant planet formation around a particular star.

Motivated by these observations, we have investigated the use of the critical or thresh-old metallicity for giant planet formation as a discriminant of different planet formation models (Rice & Armitage 2005). Using simplified core accretion models similar to those of Ida & Lin (2004a), we have calculated the radial dependence of the threshold metallicity under the assumption that disks around different stars have similar initial gas masses and lifetimes (this could be replaced with the much weaker assumption that the gas mass and disk lifetime are not correlated with the metallicity). A sample result is shown in Figure 5. By definition, planets that just manage to form as the gas disk is being lost suffer little or no Type II migration, so delineating the threshold metallicity curve observationally can yield information on planet formation that is independent of Type II migration uncertainties. We find that the most important influence on the shape of the threshold metallicity curve is probably *Type I* core migration prior to accretion of the gas envelope. When this is included, we derive a monotonically rising, minimum-metallicity curve beyond about 2 AU. In the absence of significant core migration, the threshold metallicity is flat beyond the snow line (with a weak dependence on the surface density profile of planetesimals), and the location of the snow line may be preserved in the observed distribution of planets in the orbital radius/metallicity plane.

## 5. Summary

Migration via gravitational interactions between planets and the gaseous protoplanetary disk appears to be central to understanding both the formation of gas giant planets, and their early orbital evolution to yield the extrasolar planetary systems currently being observed. Although there are uncertainties in our understanding of migration, there has also been enough progress to convince us that Type I migration is probably a vital ingredient in the formation of gas giant planets via core accretion. This is trivially true using the current best estimates of the Type I migration time scale, but it would still remain important even if the rate was suppressed by as much as two orders of magnitude. On a similarly firm footing is the assertion that gas disk migration—probably in the Type II regime—is responsible for the existence of most of the hot Jupiters. Although other migration mechanisms certainly exist, it requires moderate care to *avoid* substantial orbital evolution once a planet has formed in a gas disk.

Equally interesting are the major unknowns. Does turbulence within the disk lead to random walk migration of low-mass bodies, and if so, is this important for terrestrial planet formation and/or core accretion? Do corotation torques qualitatively change the behavior of migrating planets with masses just above the gap-opening threshold? Is the eccentricity of massive planets excited by the interaction with the gas disk? A positive answer to any of these questions would require substantial changes to our overall picture of planet formation. Addressing them will probably require, in part, high resolution simulations that include more of the complex physics of angular momentum transport within the protoplanetary disk.

This work was supported by NASA under grants NAG5-13207 and NNG04GL01G from the Origins of Solar Systems and Astrophysics Theory Programs, and by the NSF under grant AST 0407040. PJA acknowledges the hospitality of the Kavli Institute for Theoretical Physics, supported in part by the NSF under grant PHY99-07949.

### REFERENCES

ALIBERT, Y., MORDASINI, C., & BENZ, W. 2004 *A&A* **417**, L25.
ALIBERT, Y., MORDASINI, C., BENZ, W., & WINISDOERFFER, C. 2005 *A&A* **434**, 343.
ARMITAGE, P. J. 1998 *ApJ* **501**, L189.

ARMITAGE, P. J., LIVIO, M., LUBOW, S. H., & PRINGLE, J. E. 2002 *MNRAS* **334**, 248.

ARTYMOWICZ, P. 1993a *ApJ* **419**, 155.

ARTYMOWICZ, P. 1993b *ApJ* **419**, 166.

ARTYMOWICZ, P. & LUBOW, S. H. 1996 *ApJ* **467**, L77.

BALBUS, S. A. & HAWLEY, J. F. 1999 *Reviews of Modern Physics* **70**, 1.

BATE, M. R., LUBOW, S. H., OGILVIE, G. I., & MILLER, K. A. 2003 *MNRAS* **341**, 213.

BATE, M. R., OGILVIE, G. I., LUBOW, S. H., & PRINGLE, J. E. 2002 *MNRAS* **332**, 575.

BELL, K. R., CASSEN, P. M., KLAHR, H. H., & HENNING, TH. 1997 *ApJ* **486**, 372.

BODENHEIMERA, P., HUBICKYJ, O., & LISSAUER, J. J. 2000 *Icarus* **143**, 2.

BRYDEN, G., CHEN, X., LIN, D. N. C., NELSON, R. P., & PAPALOIZOU, J. C. B. 1999 *ApJ*
    **514**, 344.

CARLBERG, R. G. & SELLWOOD, J. A. 1985 *ApJ* **292**, 79.

CUMMING, A., MARCY, G. W., & BUTLER, R. P. 1999 *ApJ* **526**, 890.

D'ANGELO, G., BATE, M. R., & LUBOW, S. H. 2005 *MNRAS* **358**, 316.

D'ANGELO, G., HENNING, TH., & KLEY, W. 2002 *A&A* **385**, 647.

D'ANGELO, G., HENNING, TH., & KLEY, W. 2003 *ApJ* **599**, 548.

D'ANGELO, G., KLEY, W., & HENNING, TH. 2003 *ApJ* **586**, 540.

FISCHER, D. A. & VALENTI, J. 2005 *ApJ* **622**, 1102.

FORD, E. B., RASIO, F. A., & YU, K. 2003. In *Scientific Frontiers in Research on Extrasolar
    Planets* (eds. D. Deming & S. Seager), ASP Conference Series 294, p. 181. ASP.

FROMANG, S., TERQUEM, C., & BALBUS, S. A. 2002 *MNRAS* **329**, 18.

GAMMIE, C. F. 1996 *ApJ* **547**, 355.

GOLDREICH, P. & SARI, R. 2003 *ApJ* **585**, 1024.

GOLDREICH, P. & TREMAINE, S. 1978 *Icarus* **34**, 240.

GOLDREICH, P. & TREMAINE, S. 1979 *ApJ* **233**, 857.

GOLDREICH, P. & TREMAINE, S. 1980 *ApJ* **241**, 425.

GOODMAN, J. & RAFIKOV, R. R. 2001 *ApJ* **552**, 793.

GUILLOT, T. 2004 *Annual Review of Earth and Planetary Sciences* **33**, 493.

GULLBRING, E., HARTMANN, L., BRICENO, C., & CALVET, N. 1998 *ApJ* **492**, 323.

HAISCH, K. E., LADA, E. A., & LADA, C. J. 2001 *ApJ* **533**, L153.

HARTMANN, L., CALVET, N., GULLBRING, E., & D'ALESSIO, P. 1988 *ApJ* **495**, 385.

HAWLEY, J. F. 2001 *ApJ* **554**, 534.

HOURIGAN, K. & WARD, W. R. 1984 *Icarus* **60**, 29.

IDA, S. & LIN, D. N. C. 2004a *ApJ* **604**, 388.

IDA, S. & LIN, D. N. C. 2004b *ApJ* **616**, 567.

JANG-CONDELL, H. & SASSELOV, D. D. 2005 *ApJ* **619**, 1123.

JOHNSTONE, D., HOLLENBACH, D., & BALLY, J. 1998 *ApJ* **499**, 758.

KLEY, W., D'ANGELO, G., & HENNING, T. 2001 *ApJ* **547**, 457.

KORYCANSKY, D. G. & POLLACK, J. B. 1993 *Icarus* **102**, 150.

KUCHNER, M. J. & HOLMAN, M. J. 2003 *ApJ* **588**, 1110.

LAUGHLIN, G., STEINACKER, A., & ADAMS, F. C. 2004 *ApJ* **608**, 489.

LIN, D. N. C., BODENHEIMER, P., & RICHARDSON, D. C. 1996 *Nature* **380**, 606.

LIN, D. N. C. & IDA, S. 1997 *ApJ* **477**, 781.

LIN, D. N. C. & PAPALOIZOU, J. 1979 *MNRAS* **186**, 799.

LIN, D. N. C. & PAPALOIZOU, J. 1986 *ApJ* **309**, 846.

LISSAUER, J. J. 1993 *ARA&A* **31**, 129.

LUBOW, S. H. & OGILVIE, G. I. 1998 *ApJ* **504**, 983.

LUBOW, S. H., SEIBERT, M., & ARTYMOWICZ, P. 1999 *ApJ* **526**, 1999.

LUFKIN, G., QUINN, T., WADSLEY, J., STADEL, J., & GOVERNATO, F. 2004 *MNRAS* **347**, 421.

MARCY, G., BUTLER, R. P., FISCHER, D. A., VOGT, S. S., WRIGHT, J. T., TINNEY, C. G.,
    & JONES, H. R. A. 2005 *Progress of Theoretical Physics Sup.* **158**, 24.

MASSET, F. S. & OGILVIE, G. I. 2004 *ApJ* **615**, 1000.

MASSET, F. S. & PAPALOIZOU, J. C. B. 2003 *ApJ* **588**, 494.

MATSUMURA, S. & PUDRITZ, R. E. 2005 *ApJ* **618**, L137.

MAYOR, M. & QUELOZ, D. 1995 *Nature* **378**, 355.
MENOU, K. & GOODMAN, J. 2004 *ApJ* **606**, 520.
MIYOSHI, K., TAKEUCHI, T., TANAKA, H., & IDA, S. 1999 *ApJ* **516**, 451.
MIZUNO, H. 1980 *Progress of Theoretical Physics* **64**, 544.
MURRAY, N., HANSEN, B., HOLMAN, M., & TREMAINE, S. 1998 *Science* **279**, 69.
NELSON, A. F. & BENZ, W. 2003a *ApJ* **589**, 578.
NELSON, A. F. & BENZ, W. 2003b *ApJ* **589**, 556.
NELSON, R. P. & PAPALOIZOU, J. C. B. 2003 *MNRAS* **339**, 993.
NELSON, R. P. & PAPALOIZOU, J. C. B. 2004 *MNRAS* **350**, 849.
NELSON, R. P., PAPALOIZOU, J. C. B., MASSET, F., & KLEY, W. 2000 *MNRAS* **318**, 18.
OGILVIE, G. I. 2001 *MNRAS* **325**, 231.
OGILVIE, G. I. & LUBOW, S. H. 2003 *ApJ* **587**, 398.
PAPALOIZOU, J. C. B. 2002 *A&A* **388**, 615.
PAPALOIZOU, J. C. B. & LARWOOD, J. D. 2000 *MNRAS* **315**, 823.
PAPALOIZOU, J. C. B. & LIN, D. N. C. 1984 *ApJ* **285**, 818.
PAPALOIZOU, J. C. B., NELSON, R. P., & MASSET, F. 2001 *A&A* **366**, 263.
PAPALOIZOU, J. C. B., NELSON, R. P., & SNELLGROVE, M. D. 2004 *MNRAS* **350**, 829.
PAPALOIZOU, J. C. B. & TERQUEM, C. 1999 *ApJ* **521**, 823.
PAPALOIZOU, J. C. B. & TERQUEM, C. 2001 *MNRAS* **325**, 221.
POLLACK, J. B., HUBICKYJ, O., BODENHEIMER, P., LISSAUER, J. J., PODOLAK, M., & GREEN-ZWEIG, Y. 1996 *Icarus* **124**, 62.
PRINGLE, J. E. 1991 *MNRAS* **248**, 754.
RAFIKOV, R. R. 2002 *ApJ* **572**, 566.
RASIO, F. A. & FORD, E. B. 1996 *Science* **274**, 954.
RICE, W. K. M. & ARMITAGE, P. J. 2003 *ApJ* **598**, L55.
RICE, W. K. M. & ARMITAGE, P. J. 2005 *ApJ*, **630**, 1107.
SALMERON, R. & WARDLE, M. 2005 *MNRAS* **361**, 45.
SANO, T., MIYAMA, S. M., UMEBAYASHI, T., & NAKANO, T. 2000 *ApJ* **543**, 486.
SANTOS, N. C., ISRAELIAN, G., & MAYOR, M. 2004 *A&A* **415**, 1153.
SASSELOV, D. D. & LECAR, M. 2000 *ApJ* **528**, 995.
SCHÄFER, C., SPEITH, R., HIPP, M., & KLEY, W. 2004 *A&A* **418**, 325.
SHAKURA, N. I. & SUNYAEV, R. A. 1973 *A&A* **24**, 337.
SYER, D. & CLARKE, C. J. 1995 *MNRAS* **277**, 758.
TAKEUCHI, T., MIYAMA, S. M., & LIN, D. N. C. 1996 *ApJ* **460**, 832.
TANAKA, H. & IDA, S. 1999 *Icarus* **139**, 350.
TANAKA, H., TAKEUCHI, T., & WARD, W. R. 2002 *ApJ* **565**, 1257.
TERQUEM, C. 2003 *MNRAS* **341**, 1157.
THOMMES, E. W. 2005 *ApJ* **626**, 1033.
THOMMES, E. W., DUNCAN, M. J., & LEVISON, H. F. 1999 *Nature* **402**, 635.
TRILLING, D. E., BENZ, W., GUILLOT, T., LUNINE, J. I., HUBBARD, W. B., & BURROWS, A. 1998 *ApJ* **500**, 428.
TRILLING, D. E., LUNINE, J. I., & BENZ, W. 2002 *A&A* **394**, 241.
TSIGANIS, K., GOMES, R., MORBIDELLI, A., & LEVISON, H. F. 2005 *Nature* **435**, 459.
VERAS, D. & ARMITAGE, P. J. 2004 *MNRAS* **347**, 613.
WARD, W. R. 1988 *Icarus* **73**, 330.
WARD, W. R. 1989 *ApJ* **345**, L99.
WARD, W. R. 1997 *Icarus* **126**, 261.
WARD, W. R. & HAHN, J. M. 1995 *ApJ* **440**, L25.
WEIDENSCHILLING, S. J. 1977 *Astrophysics and Space Science* **51**, 153.
WEIDENSCHILLING, S. J. & MARZARI, F. 1996 *Nature* **384**, 619.
WINTERS, W. F., BALBUS, S. A., & HAWLEY, J. F. 2003a *MNRAS* **340**, 519.
WINTERS, W. F., BALBUS, S. A., & HAWLEY, J. F. 2003b *ApJ* **589**, 543.

# Observational constraints on dust disk lifetimes: Implications for planet formation

By LYNNE A. HILLENBRAND

California Institute of Technology, MS 105-24, Pasadena, CA 91105, USA

Thus far, our impressions regarding the evolutionary timescales for young circumstellar disks have been based on small number statistics. Over the past decade, however, in addition to precision study of individual star/disk systems, substantial observational effort has been invested in obtaining less detailed data on large numbers of objects in young star clusters. This has resulted in a plethora of information now enabling statistical studies of disk evolutionary diagnostics. Along an ordinate, one can measure disk presence or strength through indicators such as ultraviolet/blue excess or spectroscopic emission lines tracing accretion, infrared-excess tracing dust, or millimeter flux-measuring mass. Along an abscissa, one can track stellar age. While bulk trends in disk indicators versus age are evident, observational errors affecting both axes, combined with systematic errors in our understanding of stellar ages, both cloud and bias any such trends. Thus, detailed understanding of the physical processes involved in disk dissipation and of the relevant timescales remains elusive. Nevertheless, a clear effect in current data that is unlikely to be altered by data analysis improvements is the dispersion in disk lifetimes. Inner accretion disks are traced by near-infrared emission. Moderating a generally declining trend in near-infrared continuum excess and excess frequency with age over $<1$ to $8 \pm 4$ Myr, is the fact that a substantial fraction of rather young ($<1$ Myr old) stars apparently have already lost their inner accretion disks, while a significant number of rather old (8–16 Myr) stars apparently still retain them. By the age of 3–8 Myr, evidence for inner accretion disks for the vast majority of stars ($\sim90\%$) ceases to be apparent. Terrestrial zone dust is traced by mid-infrared emission where sufficient sensitivity and uniform data collection are only now being realized with data return from the *Spitzer Space Telescope*. Constraints on mid-disk dissipation and disk-clearing trends with radius are forthcoming.

## 1. Introduction

A long-standing paradigm for the formation of stars, and subsequently planets, involves the rotating collapse of a molecular cloud core to form a central proto-star surrounded by an infalling envelope and accreting disk on a timescale of $\sim10^5$ yr. Typical ages of revealed young T Tauri and Herbig Ae/Be stars are $\sim10^6$ yr. Gradual dispersal of the initially optically thick circumstellar material occurs in the early pre-main sequence phase as the system evolves through the final stages of disk accretion, which can last $\sim10^7$ yr or more in at least some well known cases (TW Hya, Hen 3−600, TWA 14—Muzerolle et al. 2000, 2001 and Alencar & Batalha 2002; PDS 66—Mamajek et al. 2002; ECha J0843.3−7905—Lawson et al. 2002; St 34—White & Hillenbrand 2005).

Physical processes occurring in younger disks include viscous accretion onto the central star, mass loss due to outflow, irradiation by the central star, ablation due to the stellar wind, turbulent mixing of material, stratification, and gradual settling of the dust towards the disk mid-plane—this last process a critical and limiting step in the path towards planet formation in the standard core accretion model (e.g., Weidenschilling et al. 1997, 2000; Pollack et al. 1996). The total disk mass decreases and the dust:gas mass ratio, assumed at least initially to be in the interstellar ratio, changes with time due to a combination of the above effects. Similarly, the dust particles are assumed to be interstellar-like in their composition and structure. Of particular interest here is the expected loss of dust opacity due to assembly of small particles into larger bodies that

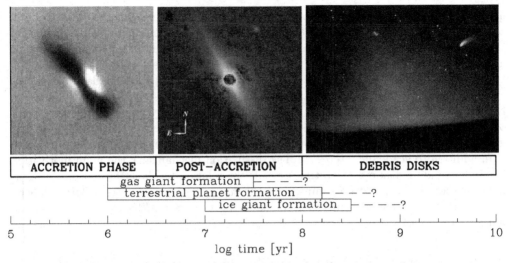

FIGURE 1. *Images of disks at various evolutionary stages scaled to a timeline showing our general understanding of the basic phenomena.* Data are courtesy of J. Stauffer and B. Patten (left panel, Ori 114−426 optically thick "silhouette disk" as imaged with *HST*/WFPC), Kalas & Jewitt (1995; middle panel, $\beta$ Pic as imaged by a ground-based coronagraph), and P. Kalas (right panel, our own zodiacal dust disk along with a comet, as photographed from Calar Alto).

might later be known as planetesimals. For solar-type stars, the ultimate result in at least 10%—and perhaps as many as 50%—of cases is a mature solar system (Marcy; this symposium).

In parallel with the discovery and study of exo-solar planets and planetary systems over the last decade (the topic of this conference), we have had dramatic observational confirmation in this same time period of the basic paradigm for star formation as briefly outlined above. Direct images and interferometric observations which spatially resolve young circumstellar disks at optical, near-infrared, and millimeter wavelengths have become common, though are far from ubiquitous. When combined with measured spectral energy distributions, such spatially resolved data are valuable for breaking model degeneracies and thus improving our understanding of source geometry and dust characteristics.

Rough correlation of the spatially resolved and SED appearances of a source, which indicate *circumstellar* status, with *stellar* evolutionary state, or age, has long been advocated (e.g., Lada 1987). However, it remains unclear whether the established sequence of circumstellar evolutionary states corresponds directly with source age. White & Hillenbrand (2004) argue for the Class I/II stages that this is not necessarily the case, given the similarities in the stellar photospheric and accretion properties of Class I and II stars as inferred from high dispersion spectroscopy of a large sample in Taurus-Auriga. Likewise, Kenyon & Hartmann (1995) discuss the Class II/III distribution in the HR diagram, which is indistinguishably intermingled, and therefore suggestive of similar ages. Because of uncertainties in age assignments, particularly for the most enshrouded sources which typically do not have ages estimated independent of their circumstellar characteristics, the timescales associated with the dispersal of circumstellar material and the formation of planets are only vaguely constrained at best.

How, then, do we catalog young circumstellar disks and characterize their evolution? Disk diagnostics come in two forms: those that trace the dust, and those that trace the gas. Dust implies small particles with typical tracers sensitive to sizes less than a mm.

These include continuum spectral energy distributions over several decades in wavelength, solid state spectroscopic features in the mid-infrared, and direct images measuring either thermal emission at long wavelengths (mid-infrared through millimeter) or scattered light at shorter wavelengths (optical and near-infrared). Gas tracers should reveal the bulk of the mass—at early stages, more than 99% of the total mass if interstellar abundances can be assumed. Sensitive gas observations of young circumstellar disks are, however, limited thus far, restricted to trace species, and dominated by upper limits. Yet recent observations of CO, $H_2$, and $H_2O$ seem promising for characterization of terrestrial zone gas. Najita (this volume) presents our knowledge of gas disk evolution in detail.

In addition to academic interest in disk dissipation mechanisms, the main motivation for understanding disk evolution timescales is the relation to planet formation. It seems prudent then to begin with a summary of the *capacity* of young disks to form planets. We will then continue with methods for assessing the *probability* that young disks do indeed form planets.

## 2. The potential for planet formation in young circumstellar disks

What are the initial conditions in young disks, and what is the likelihood that they are, in fact, proto-planetary? The raw material of planetary embryos, Earth-like rocks, and Jupiter-like gas giants is indeed abundant, if not ubiquitous, in young disks. But whether any individual disk *will* form planets is, of course, unknowable. What we can say is that many of the disks we observe are at least *capable* of forming planetary systems similar to our own, as evidenced from measured disk sizes, masses, and composition/chemistry. However, as detailed below, the *mean* disk properties are not yet known due to sensitivity limitations and therefore comparisons to our own proto-solar system based on existing data may be biased.

Disks around young stars were spatially resolved for the first time at millimeter wavelengths (e.g., Sargent & Beckwith 1987) which measure cold dust and gas in the outer disk regions. Unequal axial ratios, combined with implied dust masses large enough that the central stars should not be optically visible if the dust geometry is spherically symmetric, stood as the strongest evidence for close to a decade of disks surrounding young stars. Further, kinematic models of spatially resolved CO emission demonstrated consistency with Keplerian rotation (e.g., Koerner et al. 1993; Mannings et al. 1997; Simon et al. 2000; Qi et al. 2003).

Continued interferometric work (e.g., Lay et al. 1994; Dutrey et al. 1996; Duvert et al. 2000; Kitamura et al. 2002; Qi et al. 2003, Semenov et al. 2005), suggested that disk diameters—in instances where spatially resolved, as opposed to point-like, images are obtained—range from ~70–700 AU and are even as large as ~2000 AU in some cases. These disk-size estimates are consistent with those inferred from optical/near-infrared scattered light or silhouette images (e.g., McCaughrean & O'Dell 1996; Padgett et al. 1999; Bally et al. 2000), and in the typical case are comparable to, or larger than, the orbit of the outermost gas giant in our Solar System, Neptune. Surface density profiles, e.g., simple power-laws with $\Sigma(r) \propto r^{-p}$ or viscous disk "similarity solutions" with

$$\Sigma(r) \propto r^{-p} e^{-r^{(2-p)}} \quad ,$$

have suggested a wide range in the value of $p$ (0–1.5 for the power-law case).

Disk masses are derived from optically thin millimeter flux and an adopted opacity-wavelength relationship which leads to uncertainties of factors of 5–10 in disk masses. Under common assumptions, the calculated dust masses range from $10^{-4.5}$ to $10^{-3}$ $M_\odot$ (e.g., Beckwith et al. 1990). Making the further assumption that the dust:gas ratio by mass is unaltered from the canonical interstellar value of 1:100, total disk masses average around 0.02 $M_\odot$, or about the Minimum Mass Solar Nebula (Kusaka et al. 1970; Weidenschilling 1977), the reconstitution of present-day solar system mass and composition to solar consistency. It should be stressed that *detection at all of millimeter flux* is made amidst an increasing number of upper limits measured for stars with other indicators of disks at shorter wavelengths, and so the true "mean mass" is even lower than that quoted above.

The composition of both young primordial and older debris disks has been shown to resemble that of solar system comets. Ground-based 10 and 20 $\mu$m work on samples of brighter sources (e.g., Hanner et al. 1995, 1998; Sitko et al. 1999; van Boekel et al. 2003; Kessler-Silacci et al. 2005) and especially ISO 2–30 $\mu$m spectroscopy (e.g., Meeus et al. 2001; Bouwmann et al. 2001) have revealed an impressive suite of solid state (and PAH) dust features. Mineralogical details of the dust are modeled on a case-by-case basis due to cosmic variance, but the mean composition appears to be $\sim$70–80% amorphous magnesium-rich olivines, $\sim$1–10% crystalline forsterite, $\sim$10–15% carbons, $\sim$3–5% irons, and other trace components such as silicas. In particular, crystallinity is advocated in $\sim$10% of sources.

In summary, the *observed* sizes, masses, and chemical composition of young disks are all consistent with solar nebula estimates. This is a weak statement, however, since the *mean disk properties* are biased at present by detection limits and selection effects.

## 3. Questions concerning "primordial" dust disk evolution

The term "primordial" is used in reference to disks that are remnants of the star formation process. As outlined above, such disks are composed of the dust and gas which participated in the gravitational collapse that formed the star, and now comprise the raw materials for the formation of planets. The size, mass, and composition parameters of known young primordial disks are consistent with those estimated for the proto-solar system disk. Terrestrial planets and the rocky cores of giant planets originate from disk dust, while the gaseous envelopes of giant planets originate with the disk gas. Through either planet formation or one of the other disk dispersal mechanisms mentioned earlier, primordial disks are in the process of dissipating.

It is instructive to point out that primordial disks are physically distinguished from the so-called "debris" disks, which are secondary, rather than primordial. These gas-poor disks are comprised of dust which is regenerated during and subsequent to the growth of planets, as the large/massive bodies incite collisions amongst smaller bodies to re-form dust. Debris disks, like primordial disks, are in the process of dissipating, though via a different mechanism. Rather than sticking collisions which result in smaller particles growing to become larger particles (and eventually becoming undetectable via thermal infrared radiation), debris disk particles experience shattering collisions and gradually grind themselves down to the point at which grains are efficiently removed from the system via effects such as Poynting-Robertson drag and stellar winds. However, new dust is continually being generated in the cascade generated by collisions between the larger bodies, and the dissipating evolution is punctuated by the infusion of new material in the debris cascade.

Due to the influence of the outer giant planets on such debris, the collisional history in the inner solar system is well-documented in the cratering records on the Moon and Mars. To some degree, these records indicate the evolution of the cratering rate and the large-body size distribution with time. We have no firm record of the dust evolution in the Solar System, but even today there is "debris dust" found/assumed in the Asteroid and Kuiper belt regions. Because of the strong theoretical connection between debris dust and planetary perturbers, there is much interest in the debris belts seen around stars other than the Sun, whose evolution we can study by investigating samples of different age.

Here, I focus on the properties and evolution of dust in *primordial disks*. For any given disk, the dust mass is expected to decrease with time throughout the duration of the planet-building process, perhaps over tens of Myr. Then, if planets have successfully formed, the dust mass increases at the onset of the debris disk phase, before slowly declining again over many Gyr.

To understand the process of planet formation, we must understand how quantities such as: initial disk size and radial/vertical structure; initial disk mass and mass surface density; and initial disk composition and chemistry all evolve with time; and, further, the relative importance of various disk dispersal mechanisms (e.g., accretion, ablation, grain growth, as mentioned above). Over what timescales are dust (and gas) detectable, and how does the mass ratio of dust:gas evolve? What physical parameters determine disk longevity? What is the frequency of different end states, and in particular, that of planetary configurations? Most important for understanding the rarity or commonality of the formation of our own Solar System, what is the mean and the dispersion in all of the above distributions?

As we continue to develop the tools for answering these questions, we can also consider several pertinent "second parameter" categories. One of these relates to properties of the central star. Are there correlations in initial disk properties or disk evolution diagnostics with stellar properties such as the radiation field (particularly x-ray and ultraviolet output), stellar mass, or system metallicity, all of which may have important effects on disk structure and chemistry? A second category is related to disk physics effects. How does disk accretion history—in particular, poorly understood outburst phenomena such as FU Ori or EXOr type events—affect disk evolution? Thirdly, what is the role of environment? Multiplicity in the form of binary, triple, or quadruple systems can influence disk evolution when the companions are within, or just exterior to, the disk. Clustered versus isolated star-forming environments, in which effects such as increased ionization or photo-evaporation of disk material by massive stars, dynamical effects due to high stellar density, or the mechanical effects of multiple jets/outflows, could be important for disk evolution. Consequently, an understanding of multiplicity statistics in the form of frequency and orbital parameters, and clustering statistics in the form of spatial density and luminosity function is important for our understanding of the range of plausible disk-evolutionary paths.

In summary, there are many potentially influential parameters in the disk evolution process. The only way to effectively probe disk evolution and its many dependencies is through gathering sufficient statistics over the appropriate range of ages and "second parameter" conditions. This is a tall order indeed, but a road down which we have at least started.

## 4. Enough questions—what do we know and how do we know it?

The disk dispersal time or disk lifetime is often asserted in the literature as "about 10 Myr." This estimate is certainly good to an order of magnitude, but the justification for this number (or any other specific number) is weak at best, given the data in hand. *Some* inner dust/gas disks have disappeared within 1 Myr—by the time the star becomes optically visible. *Some* inner dust/gas disks last at least 10 Myr. In *at least one case*, that of our own solar system, some theorists have expressed the need for the gas disk to

| Property | Observational Diagnostic | Example Study |
|---|---|---|
| Disk geometry | Interferometry; SED modeling | Eisner et al. 2004, 2005b |
| Mean excess in SED; disk fraction | Broad-band photometry | Hillenbrand et al. 2006; Mamajek et al. 2004 |
| Accretion rate on to star | Ultraviolet/optical spectrophotometry | Muzerolle et al. 2000; White & Hillenbrand 2004 |
| Dust mass | Millimeter/sub-millimeter photometry | Carpenter et al. 2005; Wyatt et al. 2003 |
| Dust mineralogy; size distribution | Mid-infrared spectroscopy | Kessler-Silacci et al. 2005; van Boekel et al. 2005 |

TABLE 1. Dust disk properties measurable as a function of stellar age

survive 100 Myr or longer in order to form the outermost gas giants. As astronomers, we want to understand the mean and the dispersal time of young primordial dust/gas disks. In this section, I will review disk diagnostics, appropriate subject samples, and the difficulties involved in assessing stellar ages. In the next section, I will summarize what is known about disk evolutionary trends.

### 4.1. *Disk diagnostics*

To look for evidence of disk evolution in action, we need to consider carefully the diagnostic potential of any particular observable parameter. Many are available, however, the information obtained varies widely between different tracers of disk evolution. This is due, in part, to the variation in observational sensitivity (for example, as a function of wavelength), and in part to the varying efficacy of different disk tracers. In addition, the precision and accuracy of stellar ages—that other, often under-scrutinized (or even ignored) axis in any disk evolution diagram—needs to be critically assessed.

Deferring sensitivity considerations for the time being, what can we hope to measure as a function of stellar age? Resolved disk images, as discussed in my introductory comments, certainly have led to a wider appreciation of the convincing case for primordial "proto-planetary" disks. In fact, it is not an over-statement to say that the stunning images from ground-based interferometers (millimeter) and from the *Hubble Space Telescope* (optical and near-infrared) were responsible for transforming the field of star formation from a following of dedicated and knowledgeable disciples to high profile science. However, the reality is that few such spatially resolved images exist at present. The study of most young disk systems relies, for the most part, on so-called indirect measurements, such as broadband photometry and high resolution optical or near-infrared spectroscopy.

Table 1 details several properties of quantitative interest for young circumstellar disks and the observational diagnostics used to measure them. These are generalized properties and each can be broken down into a more detailed set of specific physical characteristics. There is an increasingly large literature on these topics, and I list only a few example studies. In most categories there is some limited evidence for at least modest evolution from primordial disk conditions. Conclusions in the area of evolution are typically based on samples of young disks ranging from small (a few) to moderate (tens to a few hundred) in size.

As mentioned above, the focus of my discussion will be on dust-disk diagnostics. In particular, I will focus on disk detection as revealed through infrared excesses—observed emission in excess of that expected from a stellar photosphere. Various levels of sophistication may be employed in the application of this technique, ranging from fully assembled spectral energy distributions covering several decades in wavelength, to two-color diagrams which cover only a limited portion of the excess spectrum, to statistical study of the disk fraction (frequency of objects in a given age bin with convincing evidence of a disk) or mean excess (magnitude or strength of the excess). Full spectral energy distributions covering ultraviolet to millimeter wavelengths have been available for only small samples of well-studied young stellar objects, making statistics difficult to assemble. Two-color diagrams are widely available, enabling statistical studies, but are more difficult to interpret without detailed knowledge of: 1) more of the spectral energy distribution, 2) the intrinsic spectral energy distribution in the absence of reddening, which can be prevalent towards young star-forming regions, and 3) the properties of the underlying star. Disk fraction and mean excess techniques account for both reddening and intrinsic stellar colors, but are based on partial spectral energy distributions. The discussion below will focus on these last techniques.

Due to several decades of ground-based work, combined with the large and uniform 2MASS photometric database at 1.2, 1.6, 2.2 $\mu$m now available, data for infrared excess investigations are abundantly available at near-infrared wavelengths. Ground-based work at mid-infrared wavelengths has been more limited in both scope and sensitivity; previous space-based platforms were revolutionary at the time, but somewhat similarly limited in sensitivity (*IRAS*) and scope (*ISO*). Thus, our understanding of disk statistics in the 3–100 $\mu$m wavelength regime is not as well developed as in the 1–2 $\mu$m regime. The *Spitzer Space Telescope* is currently accumulating sensitive data between 3–70 $\mu$m, enabling the construction of mid-infrared spectral energy distributions. These observations focus on many of the historically favored objects, though over the next few years, blind-imaging surveys of star-forming regions and young open clusters are also being conducted and will provide needed statistics.

## 4.2. *Stellar samples*

Once a technique is adopted, a sample must be chosen. In order to establish trends, robust, complete, and unbiased samples must be established over an appropriate range of ages. For the problem at hand, this includes the youngest revealed protostars through the ages characteristic of star-forming regions still associated with molecular gas (<1–2 Myr), and continuing to the entire period of terrestrial and gas giant planet formation (thought to be ~100 Myr for our own solar system), as depicted schematically in Figure 1.

Young star clusters would appear ideal for these sorts of studies because they provide the needed statistics. Furthermore, clusters have attractive attributes such as the relatively uniform distance, age, and chemical composition of their members—all of which minimize analysis complication. Young star clusters can therefore, in principle, provide the samples required to compare disk properties such as the mean and dispersion in disk lifetimes as a function of stellar mass (within a cluster), and as a function of stellar age or chemical composition (between clusters). However, careful investigation reveals that the young star samples identified to date are lacking with respect to some important issues.

First, known targets for investigations of disk evolution can be segregated into the following four coarse age groups: <1 Myr (embedded or partially embedded star forming regions), 1–3 Myr (optically revealed stellar populations still associated with molecular gas), 5–15 Myr (association members in gas-poor "fossil" star-forming regions), and finally the punctuated ages (55 Myr, 90 Myr, 120 Myr, and 600 Myr) of the nearest populous open clusters.

Of concern is that the age distribution is not uniform for known samples of young stars over the 1–100 Myr age range for disk evolution. Ample numbers (many thousands) of young stars associated with regions of recent star formation have been identified through surveys of molecular cloud complexes. Because of the intense focus on stellar census data for these young regions, such stars dominate the total numbers, and thus bias the available statistics of the overall young star age distribution towards the 1–3 Myr youngest age category.

The statistics decline dramatically at ages older than about 5 Myr and out to about 50 Myr, due to a lack of large identified samples with known ages in this "young-intermediate" age regime. There are no open clusters or large associations in the 5–50 Myr age range within 150 pc or so of the Sun, save for the Sco OB-2 association at the upper distance and lower age limit. Field stars 5–50 Myr old are extremely hard to identify, since it is only with detailed observations (not, e.g., in wide-field photometric surveys) that they stand out from much older field star populations. They may be revealed through signatures of youth, such as common proper motion with kinematically young groups, enhanced Li I absorption, Ca II H and K core emission, and x-ray activity. In fact, due to our location in the Galaxy near a ring of moderately recent star formation ("Gould's Belt"), finding stars in this age range should be relatively easy. Yet within 150 pc or so, current samples of 5–50 Myr-old stars only number a few tens, consisting of members of the TW Hya, Beta Pic, Eta Cha, and Tuc/Hor moving groups. Continued correlation of large-scale kinematic and activity databases with sufficient spectroscopic following is beginning to address this deficiency. However, the present lack of ample numbers of young stars in the 5–50 Myr age range serves to increase the error bars in disk-evolution diagnostic statistics right where the most interesting "action" of disk evolution may be taking place.

At ages older than 50 Myr, there are again ample samples due to the proximity of several nearby open clusters. Specifically, the IC2602/IC2391 pair, Alpha Per, Pleiades, and Hyades clusters, all within 200 pc and well-studied, are benchmark points in any evolutionary diagram involving either stellar or circumstellar properties.

### 4.3. *Stellar ages*

A comprehensive discussion of stellar ages is beyond the scope of this review. Suffice it to say that there are large number of age diagnostics, most of which are poorly calibrated in the young pre-main sequence age range of interest here. The most commonly used measure of stellar age in the <1–30 Myr age range is location in the Hertzsprung-Russell (HR) diagram compared to theoretical predictions of luminosity and temperature evolution as a function of time. HR diagrams (HRDs) are shown in Figure 2 for a number of current and recently star-forming regions, as well as young open clusters in the solar neighborhood.

HR diagrams can be used to infer a mean age and an apparent age dispersion as a function of effective temperature for each cluster. One issue to consider is whether the age spreads inferred for stars in young star clusters from their observed luminosity spreads indeed correspond to age ranges rather than observational errors, the default assumption. Luminosity spreads do decrease with time (e.g., consider the Orion Nebula Cluster versus the Alpha Per Cluster in Fig. 2). However, the errors in converting from observables to luminosity are the largest in the young pre-main sequence phase, just where the apparent luminosity spreads are the largest. Thus the conversion from luminosity to age—and the implied age spreads—are confusing. Understanding the age spreads, or the lack thereof, are important for the purposes of studying evolutionary diagrams. One needs either to consider all stars in a single cluster to have the mean age of the apparent distribution, or to individually assess ages and adopt an age for each star. This issue has not been satisfactorily addressed at the young ages of interest here, thus the evolutionary timescales for young circumstellar disks have large *random* uncertainties, depending on whether potentially real age spreads are accounted for in the analysis or not.

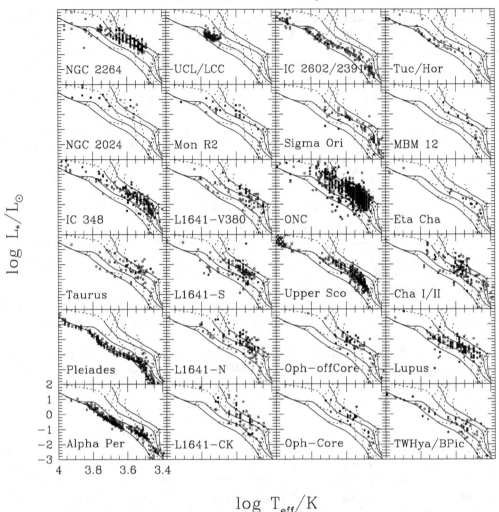

FIGURE 2. HRDs for well-studied star-forming regions and young clusters. Data were placed using the temperature scale, color scale, and bolometric corrections described in Hillenbrand & White (2004) and a wide variety of literature for the fundamental data. Pre-main sequence evolutionary calculations are those of D'Antona & Mazzitelli (1997, 1998) for isochrones of 0.1, 1.0, 10, and 100 Myr (solid lines) and masses 0.03, 0.06, 0.08, 0.1, 0.2, 0.4, 0.8, 1.5, and 3.0 $M_\odot$ (dashed lines).

Another caution is that theoretical pre-main sequence evolutionary calculations, upon which age estimates from the HR diagram rely, have significant uncertainties in their predictions. First, there is variation between various theory groups of 20–100% over certain mass and age ranges (see comprehensive discussion in Baraffe et al. 2002). Second, pre-main sequence calculations thus far do not favor well in comparison to observational constraints. Specifically, they collectively *under-predict stellar masses by 30–50%* (Hillenbrand & White 2004). Further, they *under-predict low-mass stellar ages by 30–100%* compared to lithium-depletion boundary estimates and *over-predict high-mass stellar ages by 20–100%* compared to post-main sequence evolutionary calculations. Because of this lack of theoretical validation of the age calibration of pre-main sequence isochrones, the evolutionary timescales for young circumstellar disks have large *systematic* uncertainties.

## 5. Disk evolution

I will now describe the observational constraints on the evolution of potential protoplanetary disks through the disk-clearing phase. As already emphasized, I will focus on dust disk evolution, mentioning gas where it should not be forgotten, but not discussing gas in any detail. Three spatial regimes in the disk are considered: inner disk dissipation traced by near-infrared continuum data, mid-disk dissipation depicted by mid-infrared data, and outer disk dissipation traced by millimeter wavelength data.

### 5.1. *Inner disk dissipation*

There is a well-demonstrated empirical connection between accretion and outflow diagnostics measured by high-dispersion optical spectroscopy which probes the kinematics of warm gas in the vicinity of young stars (e.g., Hartigan et al. 1995; White & Hillenbrand 2004). A similar empirical connection (e.g., Hartigan et al. 1990; Kenyon & Hartmann 1995) exists between the same spectroscopic emission lines and the blue continuum excess measured as spectroscopic veiling, both signatures of accretion directly onto the star, and photometric near-infrared (1–3 $\mu$m) continuum flux excess arising in the innermost (<0.05–0.1 AU), and thus the hottest, disk regions. These correlations affirm the basic connection between accretion from a disk and ejection in an outflow.

Furthermore, both the spectroscopic signatures of accretion and the near-infrared excess are separately demonstrated to correlate inversely with stellar age over small age ranges. Detailed modeling of the accretion temperature, density, velocity, and geometric structure is required to convert emission line strengths and profiles into mass accretion rates. More common than emission line profile studies is the measurement from high dispersion spectroscopy of continuum veiling, which can also be converted to a mass accretion rate after making assumptions about the bolometric correction to derive a total accretion luminosity, and about the infall geometry.

Treatments of the trends in the accretion rate with age have been presented by Muzerolle et al. (2000) and Calvet et al. (2005a). At least several stars appear to show measurable accretion signatures beyond 10 Myr. Existing trends have been inferred by considering the individually derived ages of stars based on their HR diagram. They are thus subject to the criticism that age spreads in individual clusters such as Taurus, Chamaeleon, or the TW Hydra association may be overestimated, and that comparisons between the mean accretion rates and mean stellar ages in each cluster may be more appropriate. Similar criticisms are also levied below against treatments of near-infrared excess behavior with age.

For the near-infrared continuum analysis, we utilize measured flux above expected photospheric values to infer disk presence. Increasingly complex inner disk geometries have been advanced over the last decade (e.g., Mahdavi & Kenyon 1998) which complicate the expectations regarding the magnitude of a near-infrared excess, given other constant parameters for the star and the disk. In the analysis discussed here, we do not consider such geometric complications and assess simply whether there is, or is not, evidence for disk emission at near-infrared wavelengths for our sample.

In calculating the color excess due to the disk, one must make two corrections from observed colors. First, it is necessary to derive and subtract the contribution from the foreground or large scale circumstellar extinction. Second, from the remaining color, a correction for the underlying stellar photosphere is performed in order to arrive at the intrinsic color excess due to the disk. In formulaic terms, using $H - K$ color as an example, the disk excess is quantified as

$$\Delta(H - K) = (H - K)_{observed} - (H - K)_{reddening} - (H - K)_{photosphere} \ .$$

Similar indices can be derived for $J - K$ or $K - L$ colors which also probe inner disk regions, though they sense dust at slightly different temperatures. In order to effect the above extinction and photospheric corrections, and hence assess intrinsic color excesses, several different sets of information are required: 1) a spectral type, for intrinsic stellar color and bolometric correction determination, 2) optical photometry, for dereddening and locating stars on the HR diagram, assuming known distance, and 3) infrared photometry, for measurement of disk "strength."

It should be borne in mind that disk strength, quantified as above from measurement of the absolute value of the infrared excess, is still a relative quantity. For any given star/disk system, the infrared excess is affected by both stellar properties (mass, radius) and disk properties (accretion rate, inclination, geometry). Meyer et al. (1997) and Hillenbrand et al. (1998; both in collaboration with Calvet) provide detailed discussions of these dependencies apropos near-infrared excesses. The effects of stellar and disk parameters on overall spectral energy distributions are discussed more comprehensively by D'Alessio et al. (1999).

Now what about that pesky other axis of stellar age? Instead of discussing in detail all of the inherent uncertainties in locating stars in the HR diagram (Fig. 2), and in inference of stellar ages and masses from those diagrams, I will simply assume fiducial cluster ages based on the median apparent age of stars in the mass range 0.3–1.0 $M_\odot$. With both a disk diagnostic and a method of cluster age estimation we can now explore the evidence for disk evolution.

Our best effort at empirically measuring the timescale for the evolution of inner circumstellar accretion disks is represented in Figure 3, which was produced from a sample of ∼3000 stars located ∼50–500 pc from the Sun. To be included in the sample, each star was required to have the spectral type, optical photometry, and infrared photometry necessary for calculation of $\Delta(H - K)$ or $\Delta(K - L)$, as described above. It should be noted that there are far fewer stars with available $L$-band photometry than available $(J)HK$ photometry. There are several important points made by these example plots.

First, even at the earliest evolutionary stages at which stars can be located in the HR diagram, the optically-thick inner disk fraction does not approach unity. There are several well-known examples of objects near the stellar birthline without any evidence for disks. This may be influenced by selection effects in that protostars and objects in transition from the protostellar to the optically revealed stage generally lack the spectroscopic data required for inclusion in our sample. However, the result is more apparent in the $H - K$ excess figure than in the $K - L$ excess figure. If a real effect (as opposed to an effect introduced by bias in the samples selected for $L$-band photometry), this strongly indicates that *some* disk evolution does happen very early on for *some* stars, before they become optically visible.

Second, beyond one Myr of age, existing samples are less biased by complications of extinction and self-embeddedness, and hence are more representative of underlying stellar populations as a whole (if not close to complete for most of the regions plotted). At these ages, there is a steady decline with time in the fraction of stars showing near-infrared excess emission (i.e., optically thick inner disks), as well as large scatter at any given age. We will return to the issue of the scatter later. The conversion of diagrams like Figure 3 into frequency distributions of accretion disk lifetimes is the next step,

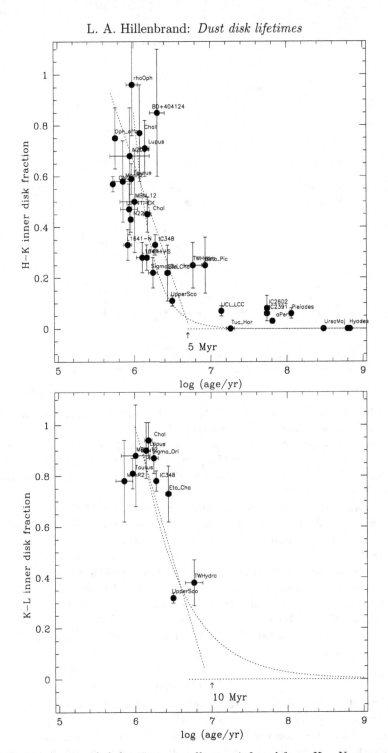

FIGURE 3. *Inner accretion disk fraction vs. stellar age* inferred from $H - K$ excess (*top panel*) and $K - L$ excess (*bottom panel*) measurements, binned by cluster or association. All young stars which we are able to locate in the HR diagram based on information in the literature (about 3500) and having inferred masses 0.3–1.0 $M_\odot$ are included in this figure. Individual clusters are treated as units of single age, corresponding to the median age inferred from the HR diagram. A cut of $\Delta(H - K) > 0.05$ mag is used to define a disk. Standard deviation of the mean (abscissa) and Poisson (ordinate) error bars are shown. The linear and exponential fits were derived for ages <30 Myr; the linear fit has a negative slope close to unity with rms 0.3.

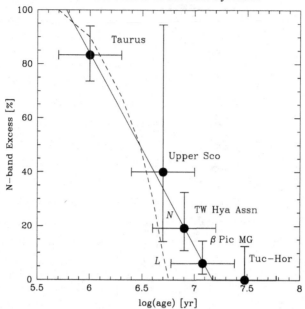

FIGURE 4. *Terrestrial zone disk fraction vs. stellar age* inferred from $N$-band excess measurements for $\sim$50 stars, taken from Mamajek et al. 2004.

and really what we want to know rather than disk frequency with age; this analysis is presented in Hillenbrand, Meyer, & Carpenter (2006).

Third, based on assessment of modern data, the median lifetime of inner optically thick accretion disks may be as short as 2–3 Myr, with essentially no evidence for $HKL$ excess present *in the median star* beyond 5 Myr. Clearly, there are exceptions such as the noted cases of $\sim$10 Myr-old accretion disks.

Other discussions of inner disk lifetimes have used different techniques and more limited samples of stars (e.g., Walter et al. 1988; Strom et al. 1989; Skrutskie et al. 1990; Beckwith et al. 1990; Strom 1995; Haisch et al. 2001). As with most scientific inquiries, the results derived depend on the details of both the samples and the analysis. Within the proposed random and systematic uncertainties, all of the above studies are comparable in their results. Previous general conclusions regarding inner disk lifetimes in the 3–10 Myr age range are, broadly speaking, similar to our findings of <2–3 Myr for the evolution of the mean disk. Further, although most disks appear to evolve relatively rapidly, a small percentage appear to retain proto-planetary nebular material for factors of 5–10 longer than the average disk.

## 5.2. *Mid-disk dissipation*

As emphasized above, near-infrared wavelengths measure hot dust in the innermost disk regions, the presence of which is well correlated empirically with independent (spectro-scopic) evidence for accretion onto the star. Because only a small amount of dust is required to make the inner disk optically thick, near-infrared continuum excesses tell us little about the bulk of the disk mass or surface area, which radiates at much cooler temperatures, and hence, longer wavelengths. Further, because the dynamical time is a function of radius in the disk, there is some expectation in the scenario that disk dissipation involves sticking collisions that eventually generate planetesimals—for disks to evolve in the inner regions first and the outer regions later (e.g., Hollenbach et al. 2000).

Thus, studying disk frequency with age (or better yet, disk lifetime) as a function of disk radius is of great interest.

Mid-infrared wavelengths, ~10–90 $\mu$m, probe disk radii ~1–5 AU, equivalent to the outer terrestrial and inner gas giant planetary zones of our solar system. To date, observational sensitivity has been the primary hindrance to measurement of disk evolution at these wavelengths. The sensitivity required at mid-infrared wavelengths is, in fact, orders of magnitude more in flux density units than that needed in the near-infrared due to the Rayleigh-Jeans fall-off of the stellar photosphere. Despite the large number of non-detections or upper limits, previous mid-infrared observations of small samples of young stars have revealed evolutionary trends.

The most recent statistical results using ground-based equipment (e.g., Mamajek et al. 2004; Metchev et al. 2004), when considered in the same excess fraction format as Figure 3, show similar morphology with ~10 Myr needed for depletion of 90% of optically thick terrestrial zone dust. Figure 4 is reproduced from Mamajek et al. 2004. The implication is that the terrestrial zone disk dissipation times are perhaps consistent with, or at most factors of a few longer than, inner disk dissipation times. If true, the combined near- and mid-infrared results suggest that disk evolution is both rapid and relatively independent of radius. However, as was true in the analysis of inner disk lifetimes, a decreasing fractional excess that is never unity is suggestive of a dispersion in disk lifetimes—in this case over an order of magnitude in age.

The *Spitzer Space Telescope* offers dramatic improvement to heretofore available mid-infrared continuum excess probes of dust evolution. *Spitzer* is sensitive to nearby *stellar photospheres between 3.5 and at least 24 $\mu$m* with additional sensitive capability out to 70 $\mu$m. *Spitzer* thus enables meaningful statistical studies of primordial (and debris—see Meyer, this volume) disk evolution within and beyond the terrestrial planet zone. Advances over the previous *IRAS/ISO* and ground capability are already revolutionizing the field. Restricting the discussion to only 8 $\mu$m results, Silverstone et al. (2006) study both field stars and cluster/association members <30 Myr old, Young et al. (2004) present results for a single 30 Myr-old cluster NGC 2547, and Stauffer et al. (2005) discuss the 120 Myr-old Pleiades cluster. All of these papers reaffirm the basic Mamajek findings that terrestrial zone dust is depleted within 10 Myr, and add needed statistics. Further *Spitzer* results are forthcoming.

At longer wavelengths, 25–60 $\mu$m, data from the *IRAS* and *ISO* satellites were even more limited in addressing disk evolution problems, again due to the sensitivity requirements of such investigations. These platforms were not capable of detecting the stellar photospheres of young stars at the necessary 150 pc distance. However, some results at 60 $\mu$m have been presented in the same form as Figure 3 (e.g., Meyer & Beckwith 2000; Robberto et al. 1999), again suggesting consistency with the Mamajek et al. (2004) results at 10 $\mu$m. However, Spangler et al. (2001) and Habing et al. (2001) argue, based on *ISO* data, for a much longer mid-infrared disk dissipation timescale, on the order of hundreds of Myr. That there may be some confusion in these two studies between primordial and debris disks as a single, continuous evolutionary path is not expected over this long timescale (see, Decin et al. 2003 for a critical assessment).

Again, *Spitzer* will revolutionize the field due to its increased sensitivity and spatial resolution over previous capabilities. Results at 24 $\mu$m for the 5 Myr-old Upper Sco association (Chen et al. 2005) and for the 10 Myr TW Hya association (Low et al. 2005) have appeared thus far. However, the mix of spectral types in these early studies relative to the roughly solar-type stars discussed above make rigorous comparisons of the disk dissipation statistics with radius premature.

### 5.3. *Outer disk dissipation*

Moving outward in wavelength and hence downward in temperature, millimeter wavelength emission probes the cold outer ($\sim$50–100 AU) disk regions and is optically thin. Most millimeter observations (e.g., Andrews & Williams 2005) have been directed towards stars younger than $\sim 10^7$ year, but because of the distance of these populations, they generally place only upper limits on dust masses beyond the youngest phase (e.g., Duvert et al. 2000). Recent application of a clever technique to push below formal detection limits has resulted in more stringent constraints on the typical disk masses in several very young clusters (Carpenter 2002; Eisner & Carpenter 2003), finding mean dust masses of $5 \times 10^{-5}$ $M_\odot$ (which can be augmented by an assumed gas-to-dust ratio to infer a total mass). Because of the more dispersed nature of older stars, there remain few such stringent constraints on typical dust masses in the 3–10 Myr-old age range.

Dust mass surveys of older ($10^7$–$10^9$ year), closer, candidate *debris* disk stars (e.g., Zuckerman & Becklin 1993; Jewitt 1994; Wyatt et al. 2003; Carpenter et al. 2005) also reveal mostly upper limits due to current sensitivity challenges, but also several detections of very proximate stars with dust masses as low as $10^{-8}$ $M_\odot$. In an analysis of the ensemble of upper limits, Carpenter et al. (2005) find marginal evidence for continuous evolution in the dust masses at expected primordial disk ages, from the 1–2 Myr young clusters to the 3–10 and 10–30 Myr field stars, which may in fact have already transitioned from primordial to debris disks.

Assessment of primordial disk evolution at radii of several tens to hundreds of AU, where the bulk of the disk mass resides, thus awaits dramatically improved millimeter and sub-millimeter sensitivity. Such is on the horizon with the commissioning of CARMA and ALMA.

## 6. How long does the dissipation process take, once initiated?

Once the process of disk dissipation starts, how long does it take for an individual object to transition from optically thick to optically thin dust? The expectation is for a short transition, based on calculations of initial grain growth via pairwise sticking collisions followed by runaway growth that forms large planetesimals (Moon-sized) on timescales of only $\sim 10^5$ yr (e.g., Wetherill & Stewart 1993; Weidenshilling & Cuzzi 1993). Is there a radial dependence to the disk clearing or do inner-, mid-, and outer-disk regimes dissipate simultaneously? While there are clear decreasing trends with advancing stellar age, both in the fraction of objects exhibiting infrared excess and in the mean magnitude of the infrared excess (not addressed in the discussion above), this does not inform us about the disk dissipation time for an individual object. The observed trends and their dispersion can be used, however, to construct statistical arguments that address the duration of the disk dissipation process, as a function of radius.

Historically, a relatively short, less than a few hundred-thousand-year timescale, has been inferred for the transition from an optically thick circumstellar disk to an optically thin circumstellar disk. The logic is based on two arguments—first, the disk statistics in binary pairs, and second, the small number (and therefore fraction) of objects found in transition between the optically thick and optically thin stages. Binary pairs, particularly in Taurus, have been well characterized in terms of the well-known CTTS (disked) and WTTS (diskless) categories. Numerous studies (e.g., Hartigan, Strom, & Strom 1994; Prato & Simon 1997; Duchene et al. 1999, Hartigan & Kenyon 2003) have found that the vast majority, >80%, of binary pairs are either both CTTS, or both WTTS, with mixed pairs relatively rare. This argues that the disk dissipation time is shorter than the absolute age difference between the members of stellar binaries.

Concerning transition objects, in the well-studied Taurus star-forming region, for example, V819 Tau and V773 Tau were argued by Skrutskie et al. (1990) to be the only two members out of approximately 150 known found with little or no near-infrared excess but small mid-infrared excess,† a result confirmed by Simon & Prato (1995) and Wolk & Walter (1996). This argument relies on the assumption of cluster coevality. As discussed above, this may not be a valid assumption at the few (2–3) Myr level. Spitzer data presented by Hartmann et al. (2005) appear to add several other objects to the "transition" category, such as CIDA 8, CIDA 11, CIDA 12, CIDA 14, DH Tau, DK TauB, and FP Tau.

Yet other Taurus objects have no evidence for excess out to 10 $\mu$m, but substantial excess at longer wavelengths. These are different from the sources detected with excess at or short-wards of 10 $\mu$m, but in transition from having optically thick to optically thin inner disks. They may be even slightly more evolved (in a circumstellar sense). One interpretation is that on the timescale that inner disk clearing has completed, these disks may be transitioning from optically thick to optically thin in the mid- or outer-disk regions. GM Aur has long been appreciated in this category (e.g., Koerner et al. 1993; Rice et al. 2003). Others with excesses only at long wavelengths were not detectable with the sensitivity of IRAS but are being revealed by Spitzer, for example CoKu Tau4 (D'Alessio et al. 2005) and DM Tau (Calvet et al. 2005b).

Collectively, both the optically thin and the inner cleared disks can be referred to as "transitional." In regions other than Taurus, the case for transitional disks has also been made. For example, Gauvin & Strom (1992) highlighted CS Cha in Chamaeleon as having a large inner cleared region (tens of AU), but a substantial far-infrared excess indicative of a robust outer disk. Nordh et al. (1996) show 7–15 $\mu$m flux ratios in Chamaeleon that are scattered around *either* the colors expected from flat/flared disks, *or* around photospheric colors, with essentially no objects located in between these groupings. These observations support the rapid transition timescales argued for Taurus members. Low et al. (2005) observed the same effect at longer wavelengths, 24 $\mu$m, in the much older TW Hya association.

In summary, in young (<3 Myr) star forming regions transition disks rare, with most stars having circumstellar material that is either consistent with an optically thick disk or not apparent at all. Further, there are specific examples of stars with dust in the terrestrial planet zone (0.1–3 AU) but not in the very inner disk (<0.05 AU). This suggests that material closest to the star may disappear first, as accretion subsides, and that the disk is cleared from the inside out. In slightly older (10 Myr) regions, the only disks left appear to be those in transition, already evolved or fully cleared in the inner disk regions, but retaining mid-infrared excesses indicative of mid-range disks.

How does disk clearing occur? While photometric studies at infrared and millimeter wavelengths such as those discussed above can provide statistics for assessing the dust disk dissipation timescale—and hence the dust disk lifetime—they tell us very little about the physics of the process. Studies of evolutionary changes in the disk structure or dust grain processes, by contrast, do provide physical insight, but are restricted to much smaller samples which can be studied in detail. Spectral energy distributions and mineralogy are two tools that can provide insight.

---

† See Duchene et al. 2003 for evidence concerning the multiplicity of V773 Tau and argument that the apparent excess can be attributed to one of the companions rather than betraying a "fossil" disk.

Typically, grain growth and disk evolution arguments have been made from measurement of the frequency dependence of continuum opacity in the expression $\tau_\nu(r) = \kappa_\nu \times \Sigma(r)$, where $\kappa_\nu \propto \nu^\beta$. The $\beta = 2$ appropriate for interstellar dust often yields in measurements of optically thin sub-/millimeter spectral energy distributions to $\beta = 0$–1 (see Miyake & Nakagawa 1993). The effects on the overall spectral energy distribution of grain growth are presented in a parameter study of disk geometry and grain properties by D'Alessio et al. (2001).

Detailed spectral energy distributions are most useful when combined with spatially resolved imaging at one or more wavelengths, enabling degeneracies in model parameters to be removed. Modeling studies of objects in different circumstellar evolutionary stages e.g., Class 0, Class I, Class II—and perhaps even Class III someday—spectral energy distributions can provide constraints on disk geometry. Some examples of such work are the analyses by Wood et al. (1998), Wolf et al. (2003), Eisner et al. (2005a), Kitamura et al. (2002), and Calvet et al. (2002). It should be borne in mind, however, that the connections between circumstellar and *stellar* evolutionary states are not yet clear.

As the dust transitions from optically thick to optically thin, perhaps as a function of radius, spectroscopy becomes an especially important tool for assessing grain size distribution and composition. Mineralogical studies reveal information about dust processing for example changes in chemical composition or mean grain size. There is evidence already for the growth of grains in young disks to sizes larger than are expected, based on the assumption that disk grain properties are consistent with those of interstellar dust. Direct probes of grain growth are spectroscopic studies that are sensitive to the opacity from particular species having particular size ranges. Work in the 8–13 $\mu$m atmospheric window (e.g., Kessler-Silacci et al. 2005, van Boekel et al. 2005) is being complemented, improved, and extended by *Spitzer* studies from 5–40 $\mu$m. Especially compelling observations would be those that can obtain spatially resolved mineralogical information. Intriguing results in this area have emerged recently from the VLTI (e.g., van Boekel et al. 2004).

## 7. Present assessment of dust clearing trends with radius

A single sentence summary of the above set of results on inner disks, mid-range disks, and outer disks is that the often quoted "10 Myr disk lifetime" is a gross generalization. While there are some clear declining trends at several wavelengths in measured disk strength and disk frequency with time, the simple fact that we can consider the quantity "disk frequency" implies that at any given age, some stars have disks while others do not, and thus a range in disk evolutionary times. The dispersion in disk lifetimes is at least factors of a few, if not an order of magnitude.

There is a some evidence that disk clearing times may be shorter in the near-infrared than in the mid-infrared, though this conclusion is not strong at present. The most conservative statement is that dust disk dissipation appears to occur within 3–8 Myr for the vast majority of stars, with minor evidence for more rapid time scales at smaller radii. The dissipation time for the mean disk may be <2–3 Myr. Sensitive observations with *Spitzer* of statistically significant samples of young stars spanning an appropriate age range are needed before such conclusions are robust, however. Such are beginning to emerge.

It should also be noted that the methods employed to date for statistical study of disks and disk lifetimes largely *detect* the presence or absence of a disk and do not tell us much about the detailed disk properties (radial/vertical structure, total mass, composition, etc.). This is another area in which the improved sensitivity and the spectroscopic capabilities of *Spitzer*, along with the spatially resolved imaging capabilities of ground-based facilities, will improve our understanding, though only for selected individual objects.

Finally, we reiterate that a complication in developing our empirical understanding of the timescales and physical processes associated with primordial disk dissipation is that soon after dusty disk material begins agglomerating to form planetesimals, the proto-planets likely collide and re-form the dust. When does a particular system go from being primordial (dominated by growth of smaller bodies into larger ones) to debris (dominated by destruction of larger bodies into smaller ones which are then removed from the system via Poynting-Robertson drag and stellar wind effects)? For disks surrounding stars with ages in the 5–15 Myr age range, there is some ambiguity as to whether they are primordial or debris disks. Several prominent examples are TW Hya, which is still accreting (Muzerolle et al. 2000), Beta Pic and AU Mic, both of which are nearby and spatially resolved, and new spatially unresolved detections in the 5–15 Myr age range emerging from *Spitzer* (e.g., Chen et al. 2005; Low et al. 2005; Silverstone et al. 2006).

As in the above discussion of primordial optically thick disks, spatial resolution is the key element for advances in debris disk studies with, for example, the color of scattered light providing critical information about the radial distribution of grain sizes (e.g., Metchev et al. 2005). Our main diagnostic for observationally distinguishing primordial disks from debris disks is the presence of gas, discussed in this proceedings in more detail by Najita.

## 8. Implications for planet formation

The discovery of exo-solar planets more than a decade ago made understanding of the connections between disks (both primordial and secondary/debris) and planets more critical than ever. The near ubiquity of circumstellar dust and gas disks around very young stars has been advocated for decades, but has been uniformly accepted by the astronomical community only within the past decade. The turning point was the availability of spatially resolved images of young gaseous and dusty disks at millimeter, sub-millimeter, infrared, and optical wavelengths. Beyond evidence for disks, the detailed information provided by 1) such images, 2) spectral energy distributions sampled over more than four decades in wavelength, and 3) dust and gas spectroscopy, is increasing our understanding of the initial conditions for planet formation. This review has concentrated on dust disk diagnostics.

However, detailed understanding of the processes of star and planet formation requires statistical assessment of global properties and evolutionary trends, in addition to the study of individual objects. Despite the large amount of data presently available, we are only now beginning to achieve the observational sensitivity needed to probe the full range of disk conditions. For the assembly of statistics, we still need to rely on traditional photometric and spectroscopic techniques, rather than well-sampled spectral energy distributions plus spatially resolved imaging at multiple well-separated wavelengths, which are available in relatively few cases.

With the statistics available at present, there are constraints on disk dissipation timescales, though they are limited in terms of the detail needed to constrain theories. Evidence for decreasing trends with age in the disk fraction, the mean disk accretion rate, and the mean disk mass are apparent. There are also signs in individual young disks of evolution from interstellar grain parameters. What may be most interesting, however, in all of these trends, is the *large dispersion about the mean at any given age*. This, in particular, speaks to the frequency distribution of paths for solar system formation and evolution.

Over the next decade, we will take the first step in understanding the possibilities for planetary formation by establishing the decay with time of primordial dust via near- and mid-infrared excess around stars of different mass. Studies to determine the timescales for dust disk dissipation should be followed by those aiming to similarly quantify timescales for gas disk dissipation. Fully constraining the time period over which the raw materials needed for planetary formation are available ultimately means following the evolution of disk surface density as a function of the radius from the central star. One outstanding problem in planning for this kind of statistically robust future is that we do not have adequate samples of stars in the 5–50 Myr age range, a critical time in planet formation and early solar system evolution.

Various theories of dust settling, planet formation, and planetary migration within disks are discussed elsewhere in these proceedings. The limited constraints from theory are consistent with the equally vague precision with which disk lifetimes can be inferred from observations of potential planetary systems now in the making. Thus the interpretation of observations is not—yet—the limiting step in solidifying our understanding of planet formation.

When, where, and how frequently do planets form in circumstellar disks? How do forming planetary systems evolve dynamically? What is the range in diversity of stable planetary system architectures? How frequent are habitable planets? How unique is our solar system? These are fairly sophisticated questions to be asking, especially given our rough knowledge of the planet formation process in our own solar system. Meteoritic evidence concerning survival time of the solar nebula suggests "several Myr" as the relevant evolutionary time scale. Studies, especially those concerning extinct radionuclides, support this time span for initial accretion, differentiation, and core formation (see e.g., the review by Wadhwa & Russell 2000). It should be emphasized that although dispersal of the solar *nebula* may occur quickly, the total duration over which inner planet formation was completed, in fact, approached 30–100 Myr.

An overarching goal of these pursuits is to connect what is observed elsewhere with the history of our own solar system, and hence enhance our appreciation of the uniqueness— or lack thereof—of it, our Earth, and in some respects the human circumstance.

## REFERENCES

ALENCAR, S. H. P. & BATALHA, C. 2002 *ApJ* **571**, 378.

ANDREWS, S. M. & WILLIAMS, J. P. 2005 *ApJ* **631**, 1134.

BALLY, J., O'DELL, C. R., & MCCAUGHREAN, M. J. 2000 *AJ* **119**, 2919.

BARAFFE, I., CHABRIER, G., ALLARD, F., & HAUSCHILDT, P. H. 2002 *A&A* **382**, 563.

BECKWITH, S. V. W., HENNING, T., & NAKAGAWA, Y. 2000. In *Protostars and Planets IV*, (eds. V. Mannings, A. P. Boss, & S. S. Russell). p. 533. Univ. Arizona Press.

BECKWITH, S. V. W., SARGENT, A. I., CHINI, R., & GUSTEN, R. 1990 *AJ* **99**, 924.

BOUWMANN, J., MEEUS, G., DE KOTER, A., HONY, S., DOMINIK, C., & WATERS, L. B. F. M. 2001 *A&A* **375**, 950.

CALVET, N., BRICEÑO, C., HERNÁNDEZ, J., HOYER, S., HARTMANN, L., SICILIA-AGUILAR, A., MEGEATH, S. T., & D'ALESSIO, P. 2005a *AJ* **129**, 935.

CALVET, N., D'ALESSIO, P., HARTMANN, L., WILNER, D., WALSH, A., & SITKO, M. 2002 *ApJ* **568**, 1008.

CALVET, N., D'ALESSIO, P., WATSON, D. M., FRANCO-HERNÁNDEZ, R., ET AL. 2005b *ApJ* **630**, 185.

CARPENTER, J. M. 2002 *AJ* **124**, 1593.

CARPENTER, J. M., WOLF, S., SCHREYER, K., LAUNHARDT, R., HENNING, T. 2005 *AJ* **129**, 1049.

CHEN, C. H., JURA, M., GORDON, K. D., & BLAYLOCK, M. 2005 *ApJ* **623**, 493.

D'ALESSIO, P., CALVET, N., & HARTMANN, L. 2001 *ApJ* **553**, 321.

D'ALESSIO, P., CALVET, N., & HARTMANN, L., LIZANO, S., CANTÓ, J. 1999 *ApJ* **527**, 893.

D'ALESSIO, P., HARTMANN, L., CALVET, N., FRANCO-HERNÁNDEZ, R., ET AL. 2005 *ApJ* **621**, 461.

D'ANTONA, F. & MAZZITELLI, I. 1997 *Mem. Soc. Astro. Ital.* **68**, 807; http://www.mporzio.astro.it/~dantona/.

DECIN, G., DOMINIK, C., WATERS, L. B. F. M., & WAELKENS, C. 2003 *ApJ* **598**, 636.

DUCHENE, G., GHEZ, A. M., MCCABE, C., & WEINBERGER, A. J. 2003 *ApJ* **592**, 288.

DUCHENE, G., MONIN, J.-L., BOUVIER, J., & MENARD, F. 1999 *A&A* **351**, 954.

DUTREY, A., GUILLOTEAU, S., DUVERT, G., PRATO, L., ET AL. 1996 *A&A* **309**, 493.

DUVERT, G., GUILLOTEAU, S., MENARD, F., SIMON, M., & DUTREY, A. 2000 *A&A* **355**, 165.

EISNER, J. A. & CARPENTER, J. M. 2003 *ApJ* **588**, 360.

EISNER, J. A., HILLENBRAND, L. A., CARPENTER, J. M., & WOLF, S. 2005a *ApJ* **635**, 396.

EISNER, J. A., HILLENBRAND, L. A., WHITE, R. J., AKESON, R. L., & SARGENT, A. I. 2005b *ApJ* **623**, 952.

EISNER, J. A., LANE, B. F., HILLENBRAND, L. A., AKESON, R. L., SARGENT, A. I. 2004 *ApJ* **613**, 1049.

GAUVIN, L. S. & STROM, K. M. 1992 *ApJ* **385**, 217.

HABING, H. J., DOMINIK, C., JOURDAIN DE MUIZON, M., LAUREIJS, R. J., ET AL. 2001 *A&A* **365**, 545.

HAISCH, K. E., LADA, E. A., & LADA, C. J. 2001 *AJ* **121**, 2065.

HANNER, M. S., BROOKE, T. Y., & TOKUNAGA, A. T. 1995 *ApJ* **438**, 250.

HANNER, M. S., BROOKE, T. Y., & TOKUNAGA, A. T. 1998 *ApJ* **502**, 871.

HARTIGAN, P., EDWARDS, S., & GHANDOUR, L. 1995 *ApJ* **452**, 736.

HARTIGAN, P., HARTMANN, L., KENYON, S. J., STROM, S. E., & SKRUTSKIE, M. F. 1990 *ApJ* **354**, 25.

HARTIGAN, P. & KENYON, S. J. 2003 *ApJ* **583**, 334.

HARTIGAN, P., STROM, K. M., & STROM, S. E. 1994 *ApJ* **427**, 961.

HARTMANN, L., MEGEATH, S. T., ALLEN, L., LUHMAN, K., CALVET, N., D'ALESSIO, P., FRANCO-HERNANDEZ, R., & FAZIO, G. 2005 *ApJ* **629**, 881.

HILLENBRAND, L. A., MEYER, M. R., & CARPENTER, J. M. 2006; in preparation.

HILLENBRAND, L. A. & WHITE, R. J. 2004 *ApJ* **604**, 741.

HOLLENBACH, D. J., YORKE, H. W., & JOHNSTONE, D. 2000. In *Protostars and Planets IV*, (eds. V. Mannings, A. P. Boss, & S. S. Russell). p. 401. Univ. Arizona Press.

JEWITT, D. C. 1994 *AJ* **108**, 661.

KALAS, P. & JEWITT, D. 1995 *AJ* **110**, 794.

KENYON, S. J. & HARTMANN, L. 1995 *ApJS* **101**, 117.

KESSLER-SILACCI, J. E., HILLENBRAND, L. A., BLAKE, G. A., & MEYER, M. R. 2005 *ApJ* **622**, 404.

KITAMURA, Y., MOMOSE, M., YOKOGAWA, S., KAWABE, R., TAMURA, M., & IDA, S. 2002 *ApJ* **581**, 357.

KOERNER, D. W., SARGENT, A. I., & BECKWITH, S. V. W. 1993 *Icarus* **106**, 2.

KUSAKA, T., NAKANO, T., & HAYASHI, C. 1970 *Prog. Theor. Phys. 44*, 1580.

LADA, C. J. 1987. In IAU Symp. 115, *Star Forming Regions* (eds. M. Peimbert & J. Jugaku). p. 1. Kluwer.

LAWSON, W. A., CRUSE, L. A.; MAMAJEK, E. E., & FEIGELSON, E. D. 2002 *MNRAS* **329**, 29.

LAY, O. P., CARLSTROM, J. E., HILLS, R. E., & PHILLIPS, T. G. 1994 *ApJ* **434**, L75.

LOW, F. J., SMITH, P. S., WERNER, M., CHEN, C., KRAUSE, V., JURA, M., & HINES, D. C. 2005 *ApJ* **631**, 1170.

MAHDAVI, A. & KENYON, S. J. 1998 *ApJ* **497**, 342.

MAMAJEK, E. E., MEYER, M. R., HEINZ, P. H., HOFFMAN, W. F., COHEN, M., & HORA, J. L. 2004 *ApJ* **612**, 496.

MAMAJEK, E. E., MEYER, M. R., & LIEBERT, J. 2002 *AJ* **124**, 1670.

MANNINGS, V., KOERNER, D. W., & SARGENT, A. I. 1997 *Nature* **388**, 555.

McCaughrean, M. J. & O'Dell, C. R. 1996 *AJ* **111**, 1977.

Meeus, G., Waters, L. B. F. M., Bouwman, J., van den Ancker, M. E., Waelkens, C., & Malfait, K. 2001 *A&A* **365**, 476.

Metchev, S. A., Eisner, J. A., Hillenbrand, L. A., & Wolf, S. 2005 *ApJ* **622**, 451.

Metchev, S. A., Hillenbrand, L. A., & Meyer, M. R. 2004 *ApJ* **600**, 435.

Meyer, M. R. & Beckwith, S. V. W. 2000. In *ISO Surveys of a Dusty Universe* (eds. D. Lemke, M. Stickle, & K. Wilke). p. 548. Springer-Verlag.

Meyer, M. R., Calvet, N., & Hillenbrand, L. A. 1997 *AJ* **114**, 288.

Miyake, K. & Nakagawa, Y. 1993 *Icarus* **106**, 20.

Muzerolle, J., Calvet, N., Briceno, C., Hartmann, L., & Hillenbrand, L. 2000 *ApJ* **535**, L47.

Muzerolle, J., Hillenbrand, L., Calvet, N., Hartmann, L., Briceño, C. 2001. In *Young Stars Near Earth: Progress and Prospects* (eds. R. Jayawardhana & T. Greene). ASP Conference Series Vol. 244, p. 245. ASP.

Nordh, L., Olofsson, G., Abergel, A., Andre, P., et al. 1996 *A&A* **315**, L185.

Padgett, D., Brandner, W., Stapelfeldt, K. R., Strom, S. E., Tereby, S., & Koerner, D. 1999 *AJ* **117**, 1490.

Pollack, J. B., Hubickyj, O., Bodenheimer, P., Lissauer, J. J., et al. 1996 *Icarus* **124**, 62.

Prato, L. & Simon, M. 1997 *ApJ* **474**, 455.

Qi, C., Kessler, J. E., Koerner, D. W., Sargent, A. I., & Blake, G. A. 2003 *ApJ* **597**, 986.

Rice, W. K. M., Wood, K., Armitage, P. J., Whitney, B. A., & Bjorkman, J. E. 2003 *MNRAS*, **342**, 79.

Robberto, M., Meyer, M. R., Natta, A., & Beckwith, S. V. W. 1999. In *The Universe as Seen by ISO* (eds. P. Cox & M. F. Kessler). p. 195. ESA-SP 427.

Sargent, A. I. & Beckwith, S. V. W. 1987 *ApJ* **323**, 294.

Sargent, A. I. & Beckwith, S. V. W. 1991 *ApJ* **382**, L31.

Semenov, D., Pavlyuchenkov, Ya., Schreyer, K., Henning, Th., Dullemond, C., Bacmann, A. 2005 *ApJ* **621**, 853.

Silverstone, M. D., Meyer, M. R., Mamajek, E. E., Hines. D. C., et al. 2006 *ApJ* **639**, 1138.

Simon, M., Dutrey, A., & Guilloteau, S. 2000 *ApJ* **545**, 1034.

Simon, M. & Prato, L. 1995 *ApJ* **450**, 824.

Sitko, M. L., Grady, C. A., Lynch, D. K., Russell, R. W., & Hanner, M. S. 1999 *ApJ* **510**, 408.

Skrutskie, M. F., Dutkevitch, D., Strom, S. E., Edwards, S., Strom, K. M., Shure, M. A. 1990 *AJ* **99**, 1187.

Spangler, C., Silverstone, M., Sargent, A. I., Becklin, E. E., & Zuckerman, B. 2001 *ApJ* **555**, 932.

Stauffer, J., Rebull, L., Carpenter, J., Hillenbrand, L., et al. 2005 *AJ* **130**, 1834.

Strom, K. M., Strom, S., Edwards, S., Cabrit, S., & Skrutskie, M. 1989 *AJ* **97**, 1451.

Strom, S. E. 1995 *RMAA* **1**, 317.

van Boekel, R., Min, M., Waters, L. B. F. M., de Koter, A., Dominik, C., van den Ancker, M. E., & Bouwman, J. 2005 *A&A* **437**, 189.

van Boekel, R., Waters, L. B. F. M., Dominik, C., Bouwman, J., de Koter, A., Dullemond, C. P., & Paresce, F. 2003 *A&A* **400**, 21.

van Boekel, R., Waters, L. B. F. M., Dominik, C., Dullemond, C. P., Tielens, A. G. G. M., & de Koter, A. 2004 *A&A* **418**, 177.

Wadhwa, M. & Russell, S. S. 2000. In *Protostars and Planets IV* (eds. V. Mannings, A. P. Boss, & S. S. Russell). p. 995. Univ. Arizona Press.

Walter, F. M., Brown, A., Mathieu, R. D., Myers, P. C., & Vrba, F. J. 1988 *AJ* **96**, 297.

Weidenschilling, S. J. 1977 *ApSS* **51**, 153.

Weidenschilling, S. J. 2000 *SSRv* **92**, 295.

WEIDENSCHILLING, S. J. & CUZZI, J. N. 1993. In *Protostars and Planets* (eds. E. H. Levy & J. I. Lunine). p. 1031. Univ. Arizona Press.

WEIDENSCHILLING, S. J., SPAUTE, D., DAVIS, D. R., MARZARI, F., OHTSUKI, K. 1997 *Icarus* **128**, 429.

WETHERILL, G. W. & STEWART, G. R. 1993 *Icarus* **106**, 190.

WHITE, R. J. & HILLENBRAND, L. A. 2004 *ApJ* **616**, 998.

WHITE, R. J. & HILLENBRAND, L. A. 2005 *ApJ* **621**, L65.

WOLF, S., PADGETT, D. L., & STAPELFELDT, K. R. 2003 *ApJ* **588**, 37.

WOLK, S. J. & WALTER, F. M. 1996 *AJ* **111**, 2066.

WOOD, K., KENYON, S. J., WHITNEY, B., & TURNBULL, M. 1998 *ApJ* **497**, 404.

WYATT, M. C., DENT, W. R. F., & GREAVES, J. S. 2003 *MNRAS* **342**, 876.

YOUNG, E. T., LADA, C. J., TEIXEIRA, P., MUZEROLLE, J., ET AL. 2004 *ApJS* **154**, 428.

ZUCKERMAN, B. & BECKLIN, E. E. 1993 *ApJ* **414**, 793.

# The evolution of gas in disks

By JOAN NAJITA

National Optical Astronomy Observatory, Tucson, AZ 85719, USA

Significant progress has been made over the last few decades in probing the gaseous component of planet-forming disks. I discuss how an understanding of the evolution of the gas in disks can help us to understand the processes of giant and terrestrial planet formation. I also discuss the observational tools that are currently available to study the gaseous component. These include *in situ* probes of the gas, as well as more indirect probes, such as stellar accretion rates. These tools can be used to probe the evolutionary status of various classes of young stars and thereby provide additional insights into the physical processes that govern planet formation.

## 1. Introduction

The discovery, ten years ago, of planets outside the solar system has transformed extrasolar planets from a topic of speculation and science fiction into one of the most compelling and rapidly advancing fields of astronomy today. As described at this conference, there are now more than 100 planets known beyond the solar system. Their orbital properties span an ever increasing range in mass, orbital radius, and eccentricity, demonstrating a diversity that is both remarkable and largely unanticipated.

The diversity in the properties of extrasolar planets has inspired multiple theories of their origins. Some of these (such as core accretion—Bodenheimer & Lin 2002; see also Hubickyj, this volume) strongly favor the creation of solar systems like our own, whereas others (such as gravitational instabilities—Boss 1997; Mayer et al. 2002; see also Durisen, this volume) are likely to lead to systems with very different planetary architectures. Thus, distinguishing between these scenarios has a strong connection to an issue of anthropic interest: the commonality (or rarity) of solar systems like our own. The possibility that multiple theories might account for the orbital properties of the extrasolar planets drives us to search beyond planetary architectures for clues to their origins, clues that may be contained in stellar metallicities, planetary atmospheres (Brown, this volume), and plausible planet formation environments, i.e., disks surrounding young stars.

## 2. Gas as a probe of giant and terrestrial planet formation

Indeed, an understanding of the evolution of the gaseous component of planet-forming disks can provide insights into the processes governing planet formation. In the context of giant planet formation, both the total disk masses (which are dominated by the gaseous component), and the lifetime of gas in the giant planet region of the disk can be used to constrain the dominant mode(s) of giant planet formation. That is, since the gravitational instability mode of giant planet formation requires disk masses that are a fair fraction of the mass of the star ($M_d \sim 0.1\ M_*$), gravitational instability may be a relatively uncommon mode of planet formation if disks are typically less massive. The gravitational instability mode also operates on very short timescales ($\ll 1$ Myr), and so is not particularly constrained by a short lifetime for the gas. In contrast, the core accretion mode of giant planet formation can make giant planets efficiently with only modest mass disks ($M_d \sim 0.01\ M_\odot$). However, it operates more leisurely, taking some 1–10 Myrs to form giant planets, so a short lifetime for the gas could limit the prevalence of this mode of giant planet formation.

The lifetime of gas in the terrestrial planet region of the disk is also of interest. This is because residual gas in this region of the disk can affect the outcome of terrestrial planet formation, i.e., the masses and eccentricities of planets, and their consequent habitability. For example, in the picture of terrestrial planet formation described by Kominami & Ida (2002), only a narrow range in residual gas column density is likely to lead to planets with Earth-like masses and eccentricities. If the gas column density in the terrestrial planet region is much larger than $1\,\mathrm{g\,cm^{-2}}$ at the epoch when lunar-mass protoplanets assemble to form terrestrial planets (at ages of a few Myr), gravitational gas drag is strong enough to circularize the orbits of the lunar-mass protoplanets, making it difficult for them to collide and build up planets with Earth-like masses, i.e., masses that would be large enough to support gaseous atmospheres. Conversely, if the gas column density in the terrestrial planet region is much less than $1\,\mathrm{g\,cm^{-2}}$, Earth-mass planets can be produced, but gravitational gas drag is too weak to circularize the orbit of the planet. As a result, only a narrow range of gas column densities $\sim 1\,\mathrm{g\,cm^{-2}}$ is expected to produce terrestrial planets with the Earth-like masses and low eccentricities that we associate with habitability on Earth.

These considerations motivate the characterization of the gaseous component of disks over a range of radii, in both the giant and terrestrial planet regions of the disk (1–20 AU) and over a range of masses—from the large gas masses characteristic of giant planet formation ($\gtrsim 1 M_J$), down to the residual gas masses of interest for the outcome of terrestrial planet formation.

## 3. Evolution of gaseous disks

From a theoretical perspective, several processes are expected to govern the evolution of gaseous disks. These include viscous accretion, which is responsible not only for the radial spreading of disks, but also the draining of the disk onto the central star. The latter effect may play an important role in the dissipation of the inner disk region (<5–10 AU). Another potentially important dissipation process is photoevaporation, where energetic photons from the central star heat the disk surface, causing it to become dynamically unbound from the system and to flow off the disk surface in a photoevaporative flow (e.g., Hollenbach et al. 2000). Photoevaporation has been suggested as an important dissipation mechanism for the outer disk region (>5–10 AU; Shu et al. 1993; Clarke et al. 2001). Finally, the planet formation process itself may play a fundamental role in dissipating gaseous disks.

From an observational perspective, little is known directly about the evolution of the gaseous component of the disk. Most of what we know comes instead from surrogate diagnostics such as emission from circumstellar dust (as measured by IR excess fraction—e.g., Haisch et al. 2001; see also Hillenbrand, this volume) and stellar accretion rates (e.g., Sicilia-Aguilar et al. 2005). Both of these show a significant decline in the 1–10 Myr range, results that are interpreted as indicating the dissipation of the gaseous disk. However, both diagnostics have important caveats associated with their use as a probe of the evolution of gaseous disks. For stellar accretion rates, we might want to know how to convert a measured stellar accretion rate into a gas column density, a topic that I will return to in Section 4. For the IR excesses, we might wonder whether the decline in IR excess as a function of age is primarily the result of disk dissipation or possibly a signature of grain growth. That is, could grain growth, an important first step in the planet formation process, render inner disks optically thin in the continuum, leaving behind a significant reservoir of gas?

Another caveat associated more specifically with *near-infrared* excess fractions as a function of age (e.g., Haisch et al. 2001) is that they probe the very inner region of the disk ≪1 AU. So any evolution that might be deduced for this region of the disk may not apply at larger radii. To address this concern, people have typically turned to the measurement of continuum excesses at longer wavelengths, at mid-infrared (e.g., Skrutskie et al. 1990; Mamajek et al. 2004; Silverstone et al. 2006) and submillimeter to millimeter (e.g., Osterloh & Beckwith 1995; Duvert et al. 2000; Andrews & Williams 2005) wavelengths. The latter have also been used to estimate disk masses. However, the possibility that the dust component of the disk experiences collisional and dynamical evolution on 1–10 Myr timescales suggests that some caution may be needed in interpreting these results.

Several physical processes beyond grain growth may alter the gas-to-dust ratio in different parts of the disk, thereby compromising the ability of the dust to track the evolution of the gaseous component in a simple way. For example, in the mid- to far-infrared, collisions between planetesimals may create a new population of small dust grains, decreasing the gas-to-dust ratio. Indeed, terrestrial planet formation is believed to involve planetesimal and protoplanet collisions at ages of 1–10 Myr—collisions which are expected to generate observable collisional debris (Kenyon & Bromley 2004). As a result, infrared excesses may be detected, even from systems with highly depleted gaseous disks. At longer wavelengths, it is well known that the submillimeter-to-millimeter spectral slopes generally indicate that grains have grown to millimeter sizes or larger at distances of ∼100 AU in Myr-old disks. Takeuchi & Lin (2005) show that gas drag is capable of causing such large grains to lose angular momentum and spiral into the star on Myr timescales. Thus, a reduction in submillimeter excess on Myr timescales may not rule out the survival of massive gaseous disks at large radii. These caveats suggest that using direct probes of the gaseous component of disks can provide useful, possibly critical information on the evolution of gaseous disks.

## 4. In situ probes of gaseous disks

Significant progress has been made over the few decades in developing *in situ* probes of the gaseous component of disks. As a result, a wide array of tools are now available for the study of gaseous disks over a wide range of disk radii. In the outer disk region (>10 AU), the available probes include millimeter molecular transitions, rovibrational transitions of $H_2$ in the near-infrared (e.g., Bary et al. 2003), and atomic lines at optical wavelengths (e.g., Brandeker et al. 2004). In the giant planet region of the disk, we hope to use mid-infrared atomic and molecular transitions. At the smallest radii, in the terrestrial planet region of the disk, the available diagnostics now include near-infrared transitions of CO, OH, and water (e.g., Najita et al. 2000 for a review), as well as transitions of $H_2$ at UV wavelengths (e.g., Herczeg et al. 2002). To illustrate the current state of affairs, I will highlight some of the successes and challenges associated with the use of these diagnostics in studying gaseous disks. Because of limited time, the discussion is representative rather than comprehensive, focusing on only one diagnostic from each region of the disk.

### 4.1. *Kuiper Belt and beyond: Millimeter molecular transitions*

Millimeter molecular transitions are the traditional probes of gas in the outer regions of circumstellar disks, and one of the most popular diagnostics are the pure rotational lines of CO (e.g., Sargent & Beckwith 1991; Dutrey et al. 1996). These transitions have been used with great success to show that the gaseous emission can be spatially resolved, typically on size scales >100 AU. Measurements of the spatially resolved velocity field of the emission further demonstrate that disks are rotating (e.g., Koerner 1995). Such measurements have been used with great success to measure both dynamical masses for the central stars, as well as system inclinations (e.g., Simon et al. 2000).

Due in part to the complex thermal-chemical structure of the disk, it has proven more difficult to estimate disk masses using these diagnostics. That is, the CO-observed emission is currently believed to arise from a warm layer at an intermediate height in the disk that is bounded from above by the photodissociation of CO, and from below by the condensation of CO in on cold grains in the disk midplane (e.g., Aikawa et al. 2002). Therefore, in order to derive a mass from the CO measurements, one needs to correct for the relative abundance of CO in the region where the emission arises, as well as to correct for the mass, primarily in the midplane, that is not directly probed by the CO emission. As a result of this complexity, disk masses have traditionally been estimated from the dust continuum emission instead (e.g., Beckwith et al. 1990).

Several surveys have been carried out using CO-rotational transitions for the purpose of studying the gas dissipation timescale in the outer region of the disk. One of the classics in this regard is the survey of Zuckerman et al. (1995), who inferred a dissipation timescale of <1 Myr based on CO transitions in a sample of young stars in the age range 1–10 Myr. At the time that the survey was carried out, the importance of depletion onto grains was just being recognized as an issue, which is therefore a caveat associated with the Zuckerman et al. study. More recent studies have continued to make progress in this area, e.g., the detection of strong CO emission from a weakly accreting young star (the T Tauri star V836 Tau—Duvert et al. 2000), and the demonstration that gas in the outer disk can survive for ~10 Myr around some Herbig AeBe stars (e.g., Dent et al. 2005). However, the limited sensitivity available to date has made it difficult to explore the commonality of CO emission from older and weakly accreting stars.

With the advent of the higher sensitivity and higher angular resolution available with ALMA, there is a bright future for resolving some of these difficulties. The use of diagnostics that specifically probe higher densities and smaller disk radii (e.g., $HCO^+$ 4–3; Greaves 2004) may allow these studies to extend down to the smaller radii (<30 AU) more directly relevant to planet formation. Finally, detailed models of the thermal-chemical structure of disks that follow the vertical as well as the radial structure will be needed to derive robust gas masses from the observations.

## 4.2. *Giant planet region: Mid-infrared transitions*

At smaller radii, we hope to use the suite of transitions accessible in the mid-infrared in order to probe the properties of the gas in the giant planet region of the disk (~1–20 AU). These transitions include atomic fine-structure transitions, as well as transitions of molecules such as water, OH, and molecular hydrogen. The molecular hydrogen transitions are of particular interest, since molecular hydrogen dominates the gas mass and the pure rotational transitions in the mid-infrared are capable of probing the ~100 K temperatures expected for this region.

A recent study by Gorti & Hollenbach (2004) illustrates the ability of the transitions in this spectral region to probe the giant planet region in optically thin (debris) disks. Further work on the emission expected for optically thick disks is in progress. The results for optically thin disks illustrate some of the possible challenges associated with using these diagnostics to estimate gas masses. In these models, many of the diagnostics arise from a complex thermal-chemical structure that varies strongly with disk radius and vertical height (e.g., Hollenbach et al. 2005)—a situation similar to that found for the outer disk region. Since many of the diagnostics, including molecular hydrogen, arise from a restricted range of radii, significant corrections are needed to convert measured masses of warm $H_2$ to total disk masses. Since the strengths of lines may depend on assumptions about the disk, such as the location of the inner disk radius (e.g., Hollenbach et al. 2005), high resolution spectra will likely be needed in order to break possible degeneracies and to reliably convert detected line strengths into disk masses.

Several surveys using the $H_2$ pure rotational transitions have attempted to probe the gas dissipation timescale in the giant planet region of the disk. One of the most thought-provoking studies was one carried out with the *Infrared Space Observatory* (*ISO*), which reported the surprising detection of approximately Jupiter-masses of gas residing in ~20-Myr-old debris disk systems (Thi et al. 2001). This intriguing result is thus far unconfirmed from the ground (Richter et al. 2002; Sheret et al. 2003; Sako et al. 2005) or with *Spitzer* (e.g., Chen et al. 2004). The ground-based observations are typically carried out with higher angular resolution than the *ISO* observations, so a possible origin for the discrepancy between the results is that *ISO* was primarily sensitive to extended $H_2$ emission not physically associated with the disk in the system.

Although the ground-based observations do not confirm the large line luminosities reported by *ISO*, $H_2$ detections have been made (e.g., with the Texas Echelon Cross Echelle Spectrograph [TEXES] on the InfraRed Telescope Facility), albeit at lower flux levels than those reported by *ISO*. When detected, the emission is generally narrow, indicating that the emission arises from disk radii beyond 10 AU (M. Richter, personal communication). Thus, searches for the weaker emission that might originate from smaller radii, in the heart of the giant planet region of the disk, requires moving TEXES to a higher sensitivity platform such as the Gemini telescope or a future ground-based 30-m telescope.

### 4.3. *Terrestrial planet region: CO fundamental emission*

At smaller disk radii, in the terrestrial planet region of the disk, the fundamental vibrational transitions of CO are a possible probe of the gaseous disk. This diagnostic is appealing in that it is relatively common—it is detected in nearly all actively accreting, young low-mass stars (i.e., classical T Tauri stars; e.g., Najita et al. 2003). Fundamental emission is also detected among the higher mass Herbig AeBe stars (Brittain et al. 2003; Blake & Boogert 2004). The line profiles of the emission seen among the classical T Tauri stars (FWHM $\sim$70 km s$^{-1}$) show that the fundamental transitions probe the region of the disk from the inner disk edge at $\sim$0.05 AU out to 1–2 AU, i.e., the terrestrial planet region of the disk. Since the emission spectrum includes lines covering a range of vibrational levels (e.g., $v$ = 1–0, 2–1, 3–2), as well as lines of $^{13}$CO, the transitions together probe a wide range of temperatures and column densities ($10^{-4} - 1$ g cm$^{-2}$; Fig. 1).

The relative strengths of the emission lines show that they arise from surprisingly warm gas ($\sim$1000 K), much warmer than the dust temperatures expected for the terrestrial planet region (<400 K at 1 AU; e.g., D'Alessio et al. 1998). Recent models of the vertical thermal-chemical structure of the terrestrial planet region of disks surrounding T Tauri stars can account for these elevated gas temperatures (Glassgold et al. 2004). While the gaseous atmosphere is heated by stellar x-rays and by the mechanical heating that arises from accretion, the dust component is heated by longer wavelength stellar photons. This surface heating induces a vertical temperature inversion in both the gas and dust components. However, at the low densities characteristic of the upper atmosphere of the disk, the poor thermal coupling between the gas and dust components allows the gas to achieve a much higher temperature than the dust. As a result, spectral features that form in the temperature inversion region will appear in emission.

Looking at the chemistry that takes place within this thermal structure, we find that the transition from atomic carbon to CO takes place at a vertical hydrogen column den-

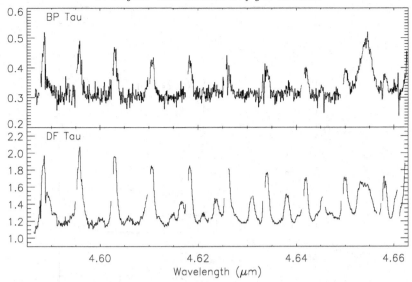

FIGURE 1. CO emission from classical T Tauri stars. Regions of strong telluric absorption have been excised from the plot. The centrally peaked line profiles show that the fundamental transitions probe a wide range of disk radii, from inner disk edge at $\sim$0.05 AU out to 1–2 AU, i.e., the terrestrial planet region of the disk. The emission spectrum can be rich (e.g., DF Tau), displaying lines from a range of vibrational levels (e.g., $v = 1$–0, 2–1, 3–2), as well as lines of $^{13}$CO.

sity of $\sim 10^{21}$ cm$^{-2}$ and a temperature of $\sim$1000 K, similar to the observed temperatures and column densities characterizing the fundamental emission. Thus, these models plausibly explain circumstances under which the emission arises. Models of this kind will be needed in order to infer quantities such as total gas column densities from observed spectra (Fig. 2).

Since the CO fundamental emission probes the low column densities that are of interest for understanding the outcome of terrestrial planet formation, we (John Carr, Bob Mathieu, and I) have been attempting to use it to probe the residual gas content of weak and transitional T Tauri stars in the terrestrial planet region of the disk. Since these systems show little infrared excess in the terrestrial planet region of the disk, we can use our observations to explore whether significant gas might remain in the inner disk even after the inner disk has become optically thin in the continuum.

Thus far, we have several detections and many upper limits (Najita 2004; Fig. 3). The detections include the T Tauri star V836 Tau, a $\sim$3-Myr-old system (Siess et al. 1999), with an H$\alpha$ equivalent width of 9 Å (Herbig & Bell 1988), and an estimated accretion rate of $\sim 4 \times 10^{-10}$ $M_\odot$ yr$^{-1}$ (Hartigan et al. 1995; Gullbring et al. 1998). CO emission is also detected from TW Hya (Fig. 3; see also Rettig et al. 2004, G. Blake personal communication), an $\sim$8-Myr-old T Tauri star with a mass accretion rate of $4 \times 10^{-10} - 2 \times 10^{-9}$ $M_\odot$ yr$^{-1}$ (Muzerolle et al. 2000; Alencar & Basri 2000). These results indicate the potential of using CO fundamental emission to probe the properties of residual gas disks.

There are several challenges to be faced in carrying out such a study. First, since stellar mass accretion rates are correlated with infrared excess (e.g., Kenyon & Hartmann 1995), detecting the presence of gas in systems with optically thin inner disks is nearly synonymous with detecting gas in systems with low accretion rates. As is known from our survey of classical T Tauri stars, the CO emission strength decreases with decreasing

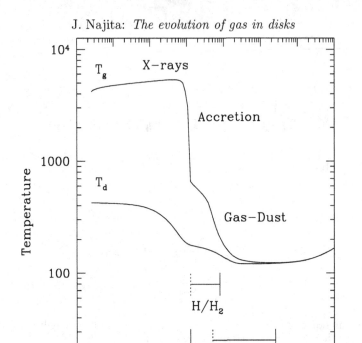

FIGURE 2. Vertical temperature structure in a disk surrounding a classical T Tauri star at a radial distance of 1 AU (Glassgold et al. 2004). Both the gas and dust components experience vertical temperature inversions, with the gas achieving much higher temperatures than the dust. In the accompanying vertical chemical structure, a modest abundance of $H_2$ $(x(H_2) = 10^{-3})$ is achieved at a vertical column density of $\sim 10^{21}$ $cm^{-2}$, which permits the conversion from atomic carbon to CO to occur at a similar column density. The gas temperature at this column density is comparable to the inferred excitation temperature of the CO fundamental emission that is detected from classical T Tauri stars. Full conversion to $H_2$ is achieved at a much larger column density $\sim 10^{22}$ $cm^{-2}$. Similarly, a modest water abundance $(x(H_2O) = 3 \times 10^{-6})$ is achieved at $\sim 5 \times 10^{21}$ $cm^{-2}$, with an asymptotic abundance $(x(H_2O) = 3 \times 10^{-4})$ achieved at much higher column densities $(>10^{23}$ $cm^{-2})$.

stellar mass accretion rate (Najita et al. 2003). This correlation might be due to the dissipation of the disk, i.e., that the total gas column density is decreasing as the accretion rate goes down. However, this seems unlikely, because the CO fundamental emission is sensitive to column densities of gas $(\sim 10^{-4}$–$1$ $cm^{-2})$ much smaller than that of primordial gas disks (100–1000 $g\,cm^{-2}$), and it seems unlikely that classical T Tauri stars have gas column densities much less than 1 $g\,cm^{-2}$. It is more likely that the decrease in emission strength among the classical T Tauri stars is due to a heating effect, that the decrease in accretion rate results in less mechanical heating for the disk atmosphere and correspondingly weaker emission, despite a large gaseous reservoir.

Emission from the weak T Tauri stars is therefore expected to be even weaker than that found for the classical T Tauri stars, so high signal-to-noise spectra are required. Secondly, since these sources also have weak continuum excesses, any emission that we might detect will be superposed on the stellar photosphere. It may therefore be necessary to correct for the spectral structure in the stellar photosphere in order to detect any weak emission. An example of this is provided by the spectrum of TW Hya (Fig. 4),

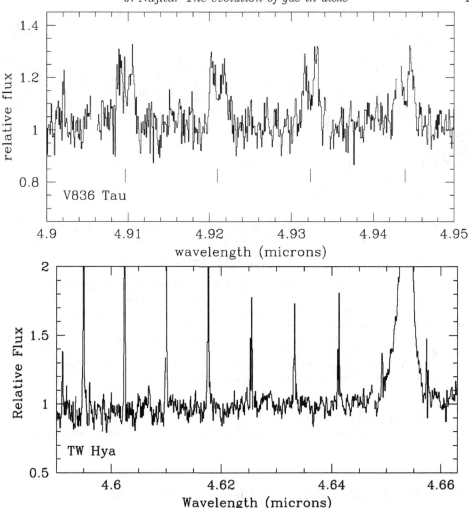

FIGURE 3. CO fundamental emission from the transitional T Tauri stars V836 Tau (top) and TW Hya (bottom), systems that are developing optically thin inner disks. The spectra show $v = 1$–$0$ emission from high-$J$ P-branch lines at 4.9 $\mu$m (top) and low-$J$ R-branch lines at 4.6 $\mu$m (bottom). Regions of strong telluric absorption have been excised from the plot. In the top panel, the vertical lines mark the approximate CO line centers at the velocity of the star. The velocity widths of the lines in the V836 Tau spectrum indicate that the emitting gas is located within a few AU of the star. For TW Hya, which is nearly face-on and therefore displays spectrally unresolved line emission, the excitation of the emitting gas suggests a similar interpretation. For V836 Tau, the relative strengths of the lines suggest optically thick emission. Thus, a large reservoir of gas may be present in the inner disk despite the weak infrared excess from this portion of the disk.

where we find that a synthetic stellar spectrum provides a good match to many of the weaker absorption features, but predicts stronger CO absorption in the low vibrational transitions than is seen in the measured spectrum. This indicates that the CO absorption in the stellar photosphere is "veiled" by CO emission from the circumstellar disk. If we were to correct for the CO absorption in the stellar photosphere, in addition to finding stronger emission in the CO $v = 1$–$0$ lines that are apparent in Figure 3, we would also uncover previously undetected emission in the higher vibrational transitions of CO. This

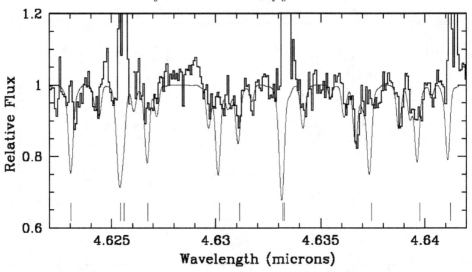

FIGURE 4. Spectrum of the transitional T Tauri star TW Hya in the 4.6 μm region (histogram), as shown in Figure 2, but here focusing on the structure in the continuum. The strong emission lines that extend above the plotted region are the bright CO fundamental emission lines seen in Figure 2. For comparison, the predicted stellar spectrum for TW Hya is also shown (lighter line). The model stellar spectrum fits many of the weaker features seen in the continuum. However, it predicts stronger stellar CO absorption in the low vibrational transitions (indicated by the lower vertical lines) than is seen in the measured spectrum. This suggests that the stellar photospheric spectrum is veiled by CO emission from warm disk gas.

would imply a higher temperature for the emitting gas that is consistent with the warm molecular gas that is inferred for this system from UV fluorescent molecular hydrogen lines (Herczeg et al. 2002).

## 5. An indirect probe of gaseous disks: Stellar accretion rates

As the above discussion suggests, the study of the evolution of gaseous disks using *in situ* diagnostics is still in its infancy, despite the significant progress that has been made. Given this situation, it may be interesting to explore the possibility of using surrogate diagnostics, such as stellar accretion rates, to probe the evolution of gaseous disks. As mentioned in Section 3, an obvious issue in doing this is how to convert stellar accretion rates into estimates of disk masses or column densities. For a steady $\alpha$ accretion disk, the disk accretion rate $\dot{M}$ is related to the disk column density $\Sigma$ at a radius $r$ through a relation of the form $\Sigma \propto \dot{M}/\alpha T$, where $T$ is the disk temperature at $r$ and the parameterized viscosity is $\nu = \alpha c_s H$. Using a relation of this form and given a measured mass accretion rate and a value for $\alpha$, we can infer a disk column density.

Before going forward, we might want to confirm that such a relation is really valid. For example, we might try to directly measure disk column densities using any of the diagnostics discussed in the previous section, and compare these with measured accretion rates in order to determine the value of $\alpha$. We might carry out suites of observations covering a range of disk radii and for several sources in order to determine if $\alpha$ is constant with radius and from source to source. Observations of this kind, which can be carried out in the future, would help to calibrate the relation between $\Sigma$ and $\dot{M}$.

It might be fun to look ahead and imagine what such a relation might imply for the evolution of gaseous disks given what we already know about stellar accretion rates as

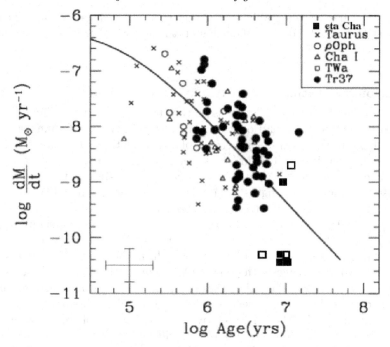

FIGURE 5. Stellar mass accretion rates as a function of age, based on data from Gullbring et al. (1998); Hartmann et al. (1998); Muzerolle et al. (1998, 2000); Lawson et al. (2004); and Sicilia-Aguilar et al. (2005). (Figure courtesy of James Muzerolle.)

a function of age (Fig. 5). If $\alpha$ is a constant as a function of radius and from source to source, the wide range in mass accretion rates at any given age (which can vary over more than an order of magnitude) imply a similarly large range in disk masses and column densities. For example, for $\alpha = 0.01$, a value that is commonly assumed in the literature, $\Sigma \simeq 100\,\mathrm{g\,cm}^{-2}$ at 1 AU for a typical T Tauri accretion rate of $\dot{M} = 10^{-8}\ M_\odot\ \mathrm{yr}^{-1}$ (D'Alessio et al. 1998); the range of stellar accretion rates at 0.5 Myr suggests that the mass column density at 1 AU can range between $\sim$30–1000 $\mathrm{g\,cm}^{-2}$ from source to source. At 5–10 Myr, the detection of stellar accretion rates as large as $10^{-8}\ M_\odot\ \mathrm{yr}^{-1}$ (e.g., Sicilia-Aguilar et al. 2005) suggest the possibility of long-lived gaseous disks in some systems.

Even at much lower accretion rates and/or older ages, the implied gas column densities would be dynamically significant. For example, for the weakly accreting T Tauri star V836 Tau at an age of 3 Myr, the measured mass accretion rate of $4 \times 10^{-10}\ M_\odot\ \mathrm{yr}^{-1}$, which is totally irrelevant for the buildup of the central stellar mass, nevertheless implies a gas column density of 4 $\mathrm{g\,cm}^{-2}$ at 1 AU. This column density is in the right range for a residual gas disk to have a favorable impact on the outcome of terrestrial planet formation. Perhaps more interesting is the case of the T Tauri star St34, which has a Li depletion age of 25 Myr (White & Hillenbrand 2005). The tiny mass accretion rate for the source of $2 \times 10^{-10}\ M_\odot\ \mathrm{yr}^{-1}$ implies a gas column density of 2 $\mathrm{g\,cm}^{-2}$ at 1 AU, again in the right range to play a role in producing planets with Earth-like masses and eccentricities. These examples illustrate the possibility that dynamically significant gaseous reservoirs persist over the age range (1–30 Myr) during which terrestrial planet formation is believed to occur. These arguments suggest that stellar accretion rates may prove to be a valuable probe of the gas content of disks. Making use of accretion rates as a diagnostic relies on being able to calibrate them against gas content. Thus it is of great interest to understand what the value of $\alpha$ is and whether it is a constant.

## 6. Evolutionary status of transitional T Tauri stars

Both the probes of gaseous disks discussed thus far—the *in situ* and indirect probes— can be used to explore the evolutionary status of certain classes of young stars in order to gain insights into the processes governing planet formation. One class of stars that is of interest is the transitional T Tauri stars. These are T Tauri stars that typically have photospheric colors at short wavelengths (optical and near-infrared) and strong contin- uum excesses at longer wavelengths. Examples of this class include TW Hya, DM Tau, GM Aur, and CoKu Tau/4 (Calvet et al. 2002; Rice et al. 2003; Bergin et al. 2004; D'Alessio et al. 2005; Calvet et al. 2005b). These systems make up a significant fraction ($\sim$10%) of T Tauri stars (e.g., Skrutskie et al. 1990; McCabe et al. 2006).

The deficit of emission at intermediate wavelengths in the spectral energy distributions (SEDs) of these sources suggests that the inner region of the disk has become optically thin in the continuum. The fact that these systems show the presence of outer disks, but not inner disks, has been taken as evidence that disks dissipate from the inside out (e.g., Skrutskie et al. 1990). The relatively small fraction of sources that are found to reside in this state is often interpreted as evidence that disks dissipate on a short timescale, compared to the mean lifetime of T Tauri stars, i.e., $<$1 Myr. However, other interpretations are possible. For example, a fraction of $\sim$10% is comparable to the fraction of nearby stars that are found to harbor giant planets. So these sources may have taken a particular evolutionary path—one that perhaps involves the formation of giant planets?

Indeed, the spectral energy distribution that is observed for transitional T Tauri stars could arise in several ways. One possibility is that a giant planet has formed in the inner disk and has dynamically cleared the inner disk region of gas and dust (e.g., Skrutskie et al. 1990; Marsh & Mahoney 1992; Rice et al. 2003). Another possibility is that the system might be in a more primitive state of planet formation. That is, perhaps plan- etesimals, protoplanets, or planetary cores have formed, thereby rendering the inner disk optically thin, but a gaseous envelope has not yet been accreted, and so a significant gaseous reservoir remains in the inner disk.

Yet another possibility is that the inner disk has been cleared of gas and dust, not by planet formation, but through the combination of viscous accretion and photoevap- oration, as described for example in the "UV Switch" model of Clarke et al. 2001. In this model, UV photons from the star photoevaporate away the outer disk, reducing its ability to resupply the material in the inner disk through viscous accretion. This even- tually causes the inner disk to decouple from the outer disk, whereupon it drains away onto the star. Thus, there are three plausible explanations for the SEDs of transitional T Tauri stars. If we could determine which of the three scenarios is valid for a sample of transitional T Tauri stars of known ages, this would provide a way of constraining the timescales for either forming planetary cores, or forming planetary cores and accreting gaseous envelopes, or for dissipating disks.

We could take a look at one particular source as a case study to see what we might learn. TW Hya is a good example. In their study of the SED for the source, Calvet et al. (2002) estimated a total column density of $32 \, \mathrm{g \, cm^{-2}}$ at 20 AU for TW Hya, based on the properties of the dust emission. If we look at the Clark et al. model, we find that the inner and outer disks decouple at $\sim$10 Myr, approximately the age of TW Hya, but at a column density at 20 AU that is only $0.1 \, \mathrm{g \, cm^{-2}}$—much smaller than the column density inferred for TW Hya. Thus, in the case of TW Hya, the outer disk appears to be far too massive for the "UV Switch" model to explain the observed structure of the SED.

What about the other two possibilities, that either giant planet or planetary core formation is responsible for the observed SED? Here, a measurement of the gas content of the inner disk might provide a useful discriminant. Several detections of gas have been made in the inner disk of this source (Section 4.3). So gas is known to be present, although models of the thermal-chemical structure of the inner disk are as yet unable to convert these measurements into total gas masses or column densities. In the interim, we might ask: What constraints can stellar accretion rates place on the gas content of the inner disk in this system?

Estimates of the stellar accretion rate for TW Hya range from $5 \times 10^{-10}$–$5 \times 10^{-9}$ $M_\odot$ yr$^{-1}$ (Muzerolle et al. 2000; Alencar & Basri 2000). If $\alpha = 0.01$, as is often assumed, this range of accretion rates corresponds to column densities of 1 AU of $\Sigma = 5$–$50\,\mathrm{g\,cm^{-2}}$. This is much less than the minimum mass solar nebula column density of $\sim 1000\,\mathrm{g\,cm^{-2}}$ at 1 AU, and consistent with the idea that the gas in the inner disk has been dissipated by the formation of a giant planet. However, a lower value of $\alpha = 0.0003$ has been invoked by Calvet et al. (2002) to explain various properties of the outer disk in the system. If the same value of $\alpha$ applies to the inner disk, the implied gas column density at 1 AU is $\Sigma = 100$–$1000\,\mathrm{g\,cm^{-2}}$, more similar to the column density in the minimum mass solar nebula, and more consistent with planetary core formation than giant planet formation. Determining which of these is the appropriate interpretation would help us to understand whether the age of the TW Hya system ($\sim 10$ Myr) is the characteristic time for the assembly of a planetary core, or the time that is needed to go a step further and accrete a gaseous envelope. Measurements of the gas content in such systems using *in situ* diagnostics can help to resolve this issue.

## 7. Evolutionary status of weak T Tauri stars

Another interesting population of young stars are weak T Tauri stars. These sources are accreting at low rates, if at all, and typically show no signs of infrared excess. They make up a significant fraction of T Tauri stars ($\sim 50\%$) and span a wide range of ages, from those comparable to those of the youngest classical T Tauri stars to those $>10$ Myr (Kenyon & Hartmann 1995). Thus, these sources show signs of neither circumstellar dust nor gas, as might be deduced from stellar accretion rates.

What might be the nature of these sources? Weak T Tauri stars may represent sources with very small initial disk masses, disks that dissipated rapidly and were possibly too low in mass to form any planets. These might, then, be failed planetary systems. They might also represent sources with very large initial disk masses, disks that went gravitationally unstable early on, thereby forming a planetary system and dissipating the disk (e.g., Mayer et al. 2002). Since the gravitational instability mode of planet formation works quickly, this scenario may plausibly explain the very youngest weak T Tauri stars.

Alternatively, weak T Tauri systems might have disks in which grain growth to millimeter sizes and the consequent strong dynamical interaction between the gas and grains in the disks has resulted in the rapid inspiral of the dust component, leaving behind a potentially large reservoir of gas. If grain growth has also led to the formation of planetesimals, then these might be systems in the process of giant planet formation.

In summary, the possibilities range the entire gamut in terms of the possible outcome of planet formation in these systems.

How might we distinguish between these possibilities? Systems that have successfully formed giant planets through gravitational instabilities could be identified by the presence

of massive, distant planetary companions. Searches for low-mass companions to weak
T Tauri stars in the >150 AU separation range detect potential companions at large
separations (>500 AU) in some weak T Tauri systems (Massarotti et al. 2005). Sensitive
companion searches at smaller radii <150 AU will help to determine whether planet
formation via gravitational instabilities is a plausible explanation for the properties of
weak T Tauri stars. To address the possibility that weak T Tauri stars are systems in
which grain growth and subsequent inspiral has occurred under the action of gas drag,
we might search for the presence of large gaseous reservoirs using any of the diagnostics
discussed in Section 4.

## 8. Summary

The study of the gaseous component of disks is still in its infancy. The past few
decades have seen marked improvement in our ability to actually probe the gaseous
component through the development of multiple spectral line diagnostics of the gas in
disks. These diagnostics now span a wide range in wavelength (from UV through radio)
and together probe disks from the inner disk edge, at a few stellar radii, out to the
distant reaches of the disk, beyond the Kuiper Belt region. Complementing these *in situ*
diagnostics is the more indirect method of using stellar accretion rates to probe the
gaseous component. As I have described, this possibly under-utilized tool already seems
to suggest that dynamically significant reservoirs of gas can survive over the timescales
needed to influence the outcome of terrestrial planet formation. Determining exactly
how much gas survives and how frequently this occurs requires the calibration of stellar
accretion rates using the *in situ* diagnostics.

While the modest rate of progress in this field might initially be seen as discouraging
(a sentiment expressed to me by more than one person at the symposium!), there is
great reason for optimism that more rapid progress is on the horizon. Many of the
spectral line probes that have been developed require high sensitivity and high spectral
resolution in order to be effective and widely applicable. Facilities meeting these criteria
(TEXES on Gemini; ALMA; and ground-based 30-m telescopes located at high, dry sites)
will become available over the next 1–10 years. In parallel with these advancements in
observing sensitivity, we will also need to develop and test the detailed thermal-chemical
models of gaseous disks that will be needed in order to interpret future observations.

Once the required observational sensitivity and thermal-chemical models are available,
it will be possible to use them to measure the gas dissipation timescale in disks over a
range of radii. These will place constraints on important planet formation issues, such
as the dominant modes of giant planet formation, the extent of giant planet migration,
and the outcome of terrestrial planet formation. These tools can further be used in combination with other observations to explore the evolutionary status of various classes of
T Tauri stars. Studies of the transitional T Tauri stars may be able to place constraints
on the timescale for forming planetary cores, accreting gaseous envelopes, or photoevap-
orating disks. Studies of weak T Tauri stars can address whether these are systems in
which planet formation has failed, been successful, or is ongoing. It is hoped that observations such as these will place useful constraints on the processes and outcomes of
planet formation.

I would like to thank my colleagues Steve Strom, John Carr, and Al Glassgold, with
whom I have discussed many of the issues reviewed in this contribution.

# REFERENCES

AIKAWA, Y., VAN ZADELHOFF, G. J., VAN DISHOECK, E. F., & HERBST, E. 2002 *AA* **386**, 622.

ALENCAR, S. H. P. & BASRI, G. 2000 *ApJ* **119**, 1881.

ANDREWS, S. M. & WILLIAMS, J. P. 2005 *ApJ* **631**, 1134.

BARY, J. S., WEINTRAUB, D. A., & KASTNER, J. H. 2003 *ApJ* **586**, 1138.

BECKWITH, S. V. W., SARGENT, A. I., CHINI, R. S., & GUESTEN, R. 1990 *AJ* **99**, 924.

BERGIN, E., ET AL. 2004 *ApJ* **614**, L133.

BLAKE, G. A. & BOOGERT, A. C. A. 2004 *ApJ* **606**, L73.

BODENHEIMER, P. & LIN, D. N. C. 2002 *Annual Review of Earth and Planetary Sciences* **30**, 113.

BOSS, A. P. 1995 *Science* **276**, 1836.

BRANDEKER, A., LISEAU, R., OLOFSSON, G., & FRIDLUND, M. 2004 *AA* **413**, 681.

BRITTAIN, S. D., RETTIG, T. W., SIMON, T., KULESA, C., DiSANTI, M. A., & DELLO RUSSO, N. 2003 *ApJ* **588**, 535.

CALVET, N., D'ALESSIO, P., HARTMANN, L, WILNER, D., WALSH, A., & SITKO, M. 2002 *ApJ* **568**, 1008.

CALVET, N., ET AL. 2005 *AJ* **129**, 935.

CALVET, N., ET AL. 2005 *ApJ* **630**, L185.

CHEN, C. H., VAN CLEVE, J. E., WATSON, D. M., HOUCK, J. R., WERNER, M. W., STAPELFELDT, K. R., FAZIO, G. G., & RIEKE, G. H. 2004 *American Astronomical Society Meeting Abstracts* **204**, 204.

CLARKE, C. J., GENDRIN, A., & SOTOMAYOR, M. 2001 *MNRAS* **328**, 485.

D'ALESSIO, P., CANTO, J., CALVET, N., & LIZANO, S. 1998 *ApJ* **500**, 411.

D'ALESSIO, P., ET AL. 2005 *ApJ* **621**, 461.

DENT, W. R. F., GREAVES, J. S., & COULSON, I. M. 2005 *MNRAS* **359**, 663.

DUTREY, A., GUILLOTEAU, S., DUVERT, G., PRATO, L., SIMON, M., SCHUSTER, K., & MÉNARD, F. 1996 *AA* **309**, 493.

DUVERT, G., GUILLOTEAU, S., MÉNARD, F., SIMON, M. & DUTREY, A. 2000 *AA* **355**, 165.

GLASSGOLD, A. E., NAJITA, J., & IGEA, J. 2004 *ApJ* **615**, 972.

GORTI, U. & HOLLENBACH, D. 2004 *ApJ* **613**, 424.

GREAVES, J. S. 2004 *MNRAS* **351**, L99.

GULLBRING, E., HARTMANN, L., BRICEÑO, C., & CALVET, N. 1998 *ApJ* **492**, 323.

HAISCH, K. E., LADA, E. A., & LADA, C. J. 2001 *ApJ* **553**, L153.

HARTIGAN, P., EDWARDS, S., & GHANDOUR, L. 1995 *ApJ* **452**, 736.

HARTMANN, L., CALVET, N., GULLBRING, E., & D'ALESSIO, P. 1998 *ApJ* **495**, 385.

HERBIG, G. H. & BELL, K. R. 1988 *Lick Obs. Publ.* 1111.

HERCZEG, G. J., LINSKY, J. L., VALENTI, J. A., JOHNS-KRULL, C. M., & WOOD, B. E. 2004 *ApJ* **572**, 310.

HOLLENBACH, D. J., YORKE, H. W., & JOHNSTONE, D. 2000 In *Protostars and Planets IV* (eds. V. Mannings, A. P. Boss, & S. S. Russell). p. 401. Univ. Arizona Press.

HOLLENBACH, D., ET AL. 2005 *ApJ* **631**, 1180.

KENYON, S. J. & BROMLEY, B. C. 2004 *ApJ* **602**, L133.

KENYON, S. J. & HARTMANN, L. 1995 *ApJS* **101**, 117.

KOERNER, D. W. 1995 *Ph.D. Thesis*, California Institute of Technology.

KOMINAMI, J. & IDA, S. 2002 *Icarus* **157**, 43.

LAWSON, W. A., LYO, A. R., & MUZEROLLE, J. 2004 *MNRAS* **351**, L39.

MAMAJEK, E. E., MEYER, M. R., HINZ, P. M., HOFFMANN, W. F., COHEN, M., & HORA, J. L. 2005 *ApJ* **612**, 496.

MARSH, K. A. & MAHONEY, M. J. 1992 *ApJ* **395**, L115.

MASSAROTTI, A., LATHAM, D. W., TORRES, G., BROWN, R. A., & OPPENHEIMER, B. D. 2005 *AJ* **129**, 2294.

MAYER, L., QUINN, T., WADSLEY, J., & STADEL, J. 2002 *Science* **298**, 1756.

McCABE, C., GHEX, A. M., PRATO, L., DUCHÊNE, G., FISHER, S., & TELESCO, C. 2006 *ApJ* **636**, 932.

MUZEROLLE, J., CALVET, N., BRICEÑO, C., HARTMANN, L., & HILLENBRAND, L. 2000 *ApJ* **535**, L47.

MUZEROLLE, J., HARTMANN, L., & CALVET, N. 1998 *AJ* **116**, 2965.

NAJITA, J. 2004 In *Star Formation in the Interstellar Medium: In Honor of David Hollenbach, Chris McKee and Frank Shu* (eds. D. Johnstone, F. C. Adams, D. N. C. Lin, D. A. Neufeld, & E. C. Ostriker). ASP Conf. Proc. 323, p. 271. ASP.

NAJITA, J., CARR, J. S., & MATHIEU, R. D. 2003 *ApJ* **589**, 931.

NAJITA, J., EDWARDS, S., BASRI, G., & CARR, J. 2000 In *Protostars and Planets IV* (eds. V. Mannings, A. P. Boss, & S. S. Russell). p. 457. Univ. Arizona Press.

OSTERLOH, M. & BECKWITH, S. V. W. 1995 *ApJ* **439**, 288.

RETTIG, T. W., HAYWOOD, J., SIMON, T., BRITTAIN, S. D., & GIBB, E. 2004 *ApJ* **616**, L163.

RICE, W. K. M., WOOD, K., ARMITAGE, P. J., WHITNEY, B. A., & BJORKMAN, J. E. 2003 *MNRAS* **342**, 79.

RICHTER, M. J., JAFFE, D. T., BLAKE, G. A., & LACY, J. H. 2002 *ApJ* **572**, L161.

SAKO, S., ET AL. 2005 *ApJ* **620**, 347.

SARGENT, A. I. & BECKWITH, S. V. W. 1991 *ApJ* **382**, L31.

SHERET, I., RAMSAY HOWAT, S. K., & DENT, W. R. F. 2003 *MNRAS* **343**, L65.

SHU, F. H., JOHNSTONE, D., & HOLLENBACH, D. 1993 *Icarus* **106**, 92.

SICILIA-AGUILAR, A., HARTMANN, L. W., HERNÁNDEZ, J., BRICEÑO, C., & CALVET, N. 2005 *AJ* **130**, 188.

SIESS, L., FORESTINI, M., & BERTOUT, C. 1999 *AA* **342**, 480.

SILVERSTONE, M. D., ET AL. 2006 *ApJ* **639**, 1138.

SIMON, M., DUTREY, A., & GUILLOTEAU, S. 2000 *ApJ* **545**, 1034.

SKRUTSKIE, M. F., DUTKEVITCH, D., STROM, S. E., EDWARDS, S., STROM, K. M., & SHURE, M. A. 1990 *AJ* **99**, 1187.

TAKEUCHI, T. & LIN, D. N. C. 2005 *ApJ* **623**, 482.

THI, W. F., VAN DISHOECK, E. F., BLAKE, G. A., VAN ZADELHOFF, G. J., HORN, J., BECKLIN, E. E., MANNINGS, V., SARGENT, A. I., VAN DEN ANCKER, M. E., NATTA, A., & KESSLER, J. 1995 *ApJ* **561**, 1074.

WHITE, R. J. & HILLENBRAND, L. A. 2005 *ApJ* **621**, L65.

ZUCKERMAN, B., FORVEILLE, T., & KASTNER, J. H. 1995 *Nature* **373**, 494.

# Planet formation

## By JACK J. LISSAUER

Space Science & Astrobiology Division, MS 245-3, NASA Ames Research Center,
Moffett Field, CA 94035, USA

Models of planetary growth are based upon data from our own Solar System, as well as observations of extrasolar planets and the circumstellar environments of young stars. Collapse of molecular cloud cores leads to central condensations (protostars) surrounded by higher specific angular momentum circumstellar disks. Planets form within such disks, and play a major role in disk evolution. Terrestrial planets are formed within disks around young stars via the accumulation of small dust grains into larger and larger bodies—until the planetary orbits become separated enough that the configuration is stable for the age of the system. Giant planets begin their growth as do terrestrial planets, but they become massive enough to accumulate substantial amounts of gas before the protoplanetary disk dissipates. A potential hazard to planetary systems is radial decay of planetary orbits, resulting from interactions between the planets and the natal disk. Massive planets can sweep up disk material in their vicinity, eject planetesimals and small planets into interstellar space or into their star, and confine disks in radius and azimuth. Small planetary bodies (asteroids and comets) can sequester solid grains for long periods of time and subsequently release them.

## 1. Introduction

The nearly planar and almost circular orbits of the planets in our Solar System argue strongly for planetary formation within flattened circumstellar disks. There is convincing observational evidence that stars form by gravitationally-induced compression of relatively dense regions within molecular clouds (Lada, Strom, & Myers 1993; André, Ward-Thompson, & Barsony 2000). Observations by Goodman et al. (1993) indicate that typical star-forming dense cores inside dark molecular clouds have specific angular momentum $> 10^{21}$ cm$^2$ s$^{-1}$. When these clouds undergo gravitational collapse, this angular momentum leads to the formation of pressure-supported protostars surrounded by rotationally supported disks. Such disks are analogous to the primordial solar nebula that was initially conceived by Kant and Laplace to explain the observed properties of our Solar System (e.g., Cassen et al. 1985). Observational evidence for the presence of disks of Solar System dimensions around pre-main sequence stars has increased substantially in recent years (McCaughrean, Stepelfeldt, & Close 2000). The existence of disks on scales of a few tens of astronomical units is inferred from the power-law spectral energy distribution in the infrared over more than two orders of magnitude in wavelength (Chiang & Goldreich 1997). Observations of infrared excesses in the spectra of young stars suggest that the lifetimes of protoplanetary disks span the range of $10^6$–$10^7$ years (Strom, Edwards, & Skrutskie 1993; Alencar & Batalha 2002).

Our understanding of planet formation, such as it is, comes from a diverse set of observations, laboratory studies and theoretical models. Detailed observations are now available for the planets and many smaller bodies (moons, asteroids and comets) within our Solar System. Studies of meteorite composition, minerals and physical structure have been used to deduce conditions within the protoplanetary disk (Hewins, Jones, & Scott 1996). Data on extrasolar planets are less detailed and more biased, yet still very important. Moreover, our Solar System includes the intrinsic bias of containing a habitat where life can evolve to the point of asking questions about other worlds (Wetherill 1994). Observations of young stars and their surrounding disks provide clues to the planet formation now taking place within our galaxy. Laboratory experiments on the behavior of hydrogen and helium at high pressures have been combined with gravitational measures of the mass distribution within giant planets deduced from the orbits of natural satellites and the trajectories of passing spacecraft to constrain the internal structure and composition of the largest planets within our Solar System.

Theorists have attempted to assemble all of these pieces of information together into a coherent model of planetary growth. But note that planets and planetary systems are an extremely heterogeneous lot, the 'initial conditions' for star and planet formation vary greatly within our galaxy (MacLow & Klessen 2004), and at least some aspects of the process of planet formation are extremely sensitive to miniscule changes in initial conditions (Chambers et al. 2002).

The remainder of this chapter is organized as follows: Observations are summarized in Section 2. Accretion models for terrestrial planets around single stars are reviewed in Section 3, and simulations of terrestrial planet formation around binary stars are summarized in Section 4. Giant planet formation models are discussed in Section 5, with conclusions presented in Section 6.

## 2. Planets in our Solar System and beyond

The Earth, as well as all smaller bodies within the Solar System, consists almost entirely of compounds that are condensable under reasonable conditions. In contrast, more massive planets contain a considerable fraction of light gases. About 90% of Jupiter's mass is H and He, and these two light elements make up $\sim$75% of Saturn. The large amounts of H and He contained in Jupiter and Saturn imply that these planets must have formed within $\sim$10$^7$ years of the collapse of the Solar System's natal cloud, before the gas in the protoplanetary disk was swept away. The two largest planets in our Solar System are generally referred to as *gas giants*, even though these elements aren't gases at the high pressures that most of the material in Jupiter and Saturn is subjected to. Analogously, Uranus and Neptune are frequently referred to as *ice giants*, even though the astrophysical ices (such as $H_2O$, $CH_4$, $H_2S$, and $NH_3$) that models suggest make up the majority of their mass (Hubbard, Podolak, & Stevenson 1995) are in fluid, rather than solid, form. Note that whereas H and He *must* make up the bulk of Jupiter and Saturn because no other elements can have such low densities at plausible temperatures, it is possible that Uranus and Neptune are primarily composed of a mixture of 'rock' and H/He.

Lithium and heavier elements constitute <2% of the mass of a solar composition mixture. The *atmospheric* abundances of volatile gases heavier than helium (excluding neon, which was predicted prior to Galileo Probe measurements to be substantially depleted through gravitationally induced settling; Roulston & Stevenson 1995) are $\sim$3$\times$ solar in Jupiter (Young 2003), a bit more enriched in Saturn, and substantially more for Uranus and Neptune. The *bulk* enhancements in heavy elements relative to the solar value are roughly 5, 15, and 300 times for Jupiter, Saturn, and Uranus/Neptune, respectively. Thus, all four giant planets accreted solid material substantially more effectively than gas from the surrounding nebula. Moreover, the total mass in heavy elements varies by only a factor of a few among the four planets, while the mass of H and He varies by about two orders of magnitude between Jupiter and Uranus/Neptune.

The extrasolar planet discoveries of the past decade have vastly expanded our database by increasing the number of planets known by more than an order of magnitude. The distribution of known extrasolar planets is highly biased towards those planets that are most easily detectable using the Doppler radial velocity technique, which has been by far the most effective method of discovering exoplanets. These extrasolar planetary systems are quite different from our Solar System; however, it is not yet known whether our planetary system is the norm, quite atypical, or somewhere in between.

Nonetheless, some unbiased statistical information can be distilled from available exoplanet data (Marcy et al. 2004, 2005): Roughly 1% of Sun-like stars (chromospherically quiet, late F, G and early K spectral class main sequence stars) have planets more massive than Saturn within 0.1 AU. Approximately 7% of Sun-like stars have planets more massive than Jupiter within 3 AU. Planets orbiting interior to ∼0.1 AU, a region where tidal circularization timescales are less than stellar ages, have small orbital eccentricities. The median eccentricity observed for planets on more distant orbits is 0.25, and some of these planets travel on very eccentric orbits. Within 5 AU of Sun-like stars, Jupiter-mass planets are more common than planets of several Jupiter masses, and substellar companions that are more than ten times as massive as Jupiter are rare. Stars with higher metallicity are much more likely to host detectable planets than are metal-poor stars (Gonzalez 2003; Santos et al. 2003; Fischer & Valenti 2005), with the probability of hosting an observable planet varying as the square of stellar metallicity (Marcy et al. 2005). Low-mass main-sequence stars (M dwarfs) are significantly less likely to host one or more giant planets with orbital period(s) of less than a decade than are Sun-like stars. Multiple-planet systems are more common than if detectable planets were randomly assigned to stars. Two of the three extrasolar giant planets with well-measured masses and radii, HD 209458b and TrES-1, are predominantly hydrogen (Charbonneau et al. 2000; Burrows, Sudarsky, & Hubbard 2003; Alonso et al. 2004), as are Jupiter and Saturn. However, the third such planet to be observed, HD 149026b, which is slightly more massive than Saturn, appears to have comparable amounts of hydrogen and helium and heavy elements (Sato et al. 2005), making it intermediate between Saturn and Uranus in terms of bulk composition, but more richly endowed in terms of total amount of 'metals' than is any planet in our Solar System.

Transit observations have also yielded an important negative result: *Hubble Space Telescope* photometry of a large number of stars in the globular cluster 47 Tucanae failed to detect any transiting inner giant planets, even though ∼17 such transiting objects would be expected if the frequency of such planets in this low metallicity cluster were the same as that for Sun-like stars in the solar neighborhood (Gilliland et al. 2000). In contrast, it appears likely that a ∼3 $M_J$ planet is orbiting ∼20 AU from the pulsar PSR B1620–26 white dwarf binary system, which is located in the globular cluster Messier 4. This system has been taken to be evidence for ancient planet formation in a low metallicity (5% solar) protoplanetary disk by Sigurdsson (1993) and Sigurdsson et al. (2003). Sigurdsson's formation scenario requires a fairly complex stellar exchange to account for the planet in its current orbit. There is a much more likely explanation for the planet orbiting PSR B1620–26, which requires neither planetary formation in a low metallicity disk, nor stellar exchange. This system has two post-main sequence stars sufficiently close to have undergone disk-producing mass transfer during the white dwarf's distended red giant phase, which occurred within the past $10^9$ years (Sigurdsson et al. 2003). Such a metals-enriched disk could have been an excellent location for the giant planet to form, and growth within such a disk would fit well with both planet formation theories and the observed strong correlation of planetary detections with stellar (and presumably protostellar disk) metallicity. Sigurdsson (1993) noted the possibility that the planet formed in a post-main sequence disk, but he discounted this scenario because he relied on the planetary growth timescales given by Nakano (1987), whose model requires an implausibly long $4 \times 10^9$ years to form the most distant known major planet within our Solar System, Neptune.

## 3.  Models of terrestrial planet growth around single stars

The 'planetesimal hypothesis' states that planets grow within circumstellar disks via pairwise accretion of small solid bodies known as planetesimals (Chamberlin 1905; Safronov 1969; Hayashi et al. 1977). The process of planetary growth is generally divided for convenience and tractability into several distinct stages. Planets begin to grow when microscopic dust grains collide and agglomerate via sticking/local electromagnetic forces as they settle towards the midplane of the disk. The least well-understood phase of solid planet formation is the agglomeration from cm-sized pebbles to km-sized bodies that are referred to as planetesimals. The gaseous component of the protoplanetary disk plays an important role in this stage of planetary growth (Adachi et al. 1976; Weidenschilling 1977). Collective gravitational instabilities among the solid grains in the disk (Safronov 1969; Goldreich & Ward 1973) might be important, although turbulence of the gas may prevent the protoplanetary dust layer from becoming thin enough to be gravitationally unstable (Weidenschilling & Cuzzi 1993). Recent calculations suggest that high-metallicity disks may form planetesimals via gravitational instabilities, but that dust in disks with fewer solids may not be able to overcome turbulence and settle into a subdisk that is dense enough to undergo gravitational instability (Youdin & Shu 2002). Planetesimal formation is a very active research area (Goodman & Pindor 2000; Ward 2000; Cuzzi et al. 2001), and results may have implications for our estimates of the abundance of both terrestrial and giant planets within our galaxy.

Once solid bodies reach kilometer-size (in the case of the terrestrial region of the protosolar disk), gravitational interactions between pairs of solid planetesimals provide the dominant perturbation of their basic Keplerian orbits. Electromagnetic forces, collective gravitational effects, and in most circumstances, gas drag, play minor roles. Planetesimals agglomerate via pairwise mergers. The rate of solid body accretion by a planetesimal or planetary embryo is determined by the size and mass of the planetesimal/planetary embryo, the surface density of planetesimals, and the distribution of planetesimal velocities relative to the accreting body. The evolution of the planetesimal size distribution is determined by the gravitationally enhanced collision cross-section, which favors collisions between bodies having smaller relative velocities. Runaway growth of the largest planetesimal in each accretion zone appears to be a likely outcome. The subsequent accumulation of the resulting planetary embryos leads to a large degree of radial mixing in the terrestrial planet region, with giant impacts probable. Growth via binary collisions proceeds until the spacing of planetary orbits becomes sufficient for the configuration to be stable to gravitational interactions among the planets for the lifetime of the system (Safronov 1969; Wetherill 1990; Lissauer 1993, 1995; Chambers 2001; Laskar 2000).

Safronov (1969) developed analytical tools to predict planetary growth from small ($\sim$1 km) solid bodies, although he was required to make numerous physical assumptions/approximations, as his theory did not incorporate computer modeling. Numerous groups (e.g., Greenberg et al. 1978; Wetherill & Stewart 1989; Kolvoord & Greenberg 1992) have attempted to examine the accumulation and dynamics of 1 km planetesimals via numerical simulations, using different sets of approximations to make the problem numerically tractable. Recent advances in computer hardware and coding are allowing the direct modeling of these early phases of planetary growth within localized regions of the disk (Barnes 2004; Barnes et al. 2006). Assuming perfect accretion, i.e., that all physical collisions are completely elastic, the initial stages of growth are quite rapid, especially in the inner regions of a protoplanetary disk, and large particles form quickly (Fig. 1).

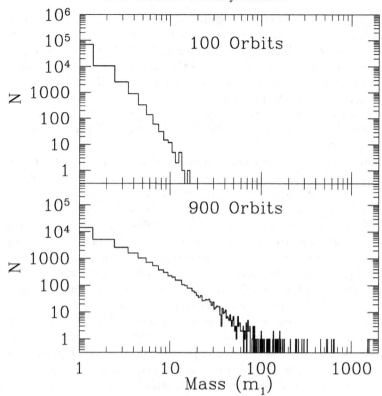

FIGURE 1. Growth of planetesimals within an accreting shearing patch of the protoplanetary disk that initially contained 106,130 planetesimals, each of radius 1 km, density 3 g/cc and mass $m_1 = 1.2 \times 10^{16}$ g, located 0.4 AU from a 1 $M_\odot$ star. The mass distribution is shown after 100 orbits (top panel) and after 900 orbits (bottom panel). After 100 orbits, all of the particles are in a distribution described by a power law. At 900 orbits, most particles are still in this type of distribution, but several particles appear to have broken away. The two most massive particles are very close in size, 1566 $m_1$ and 1518 $m_1$. Courtesy R. Barnes; see Barnes et al. (2006) for details.

Planetesimal growth regimes are sometimes characterized as either orderly or runaway. In orderly growth, the particles containing the most mass double their masses in about the same amount of time as the largest particle. The stochasticity of the system creates a small fraction of larger particles. Runaway growth involves a qualitative distinction between the particles of the system. When the relative velocity between planetesimals is comparable to or larger than the escape velocity, $v \gtrsim v_e$, the growth rate is approximately proportional to $R^2$, where $R$ is the radius of the growing planetesimal, and the evolutionary path of the planetesimals exhibits an orderly growth of the entire size distribution. When the relative velocity is small, $v \ll v_e$, the growth rate is proportional to $R^4$. In this situation, the planetary embryo rapidly grows much larger than any other planetesimal in its accretion zone, which can lead to runaway growth. By virtue of its large, gravitationally enhanced cross section, a runaway particle doubles its mass faster than smaller bodies, and detaches itself from the mass distribution (Wetherill & Stewart 1989; Ohtsuki, Stewart, & Ida 2002).

Local models break down when the impinging velocities no longer result from a homogeneous distribution. This occurs when a particle transitions from dispersion-dominated growth to shear-dominated growth (Lissauer 1987). Initially, growth is dispersion dominated; the velocity distribution of incoming particles is simply that of the velocity dispersion of the planetesimal swarm. However, when a particle reaches a size such that the Keplerian shear across its Hill sphere (the volume of space over which the body's gravity dominates over the tidal force resulting from the gradient of the star's gravitational potential) is larger than the velocity dispersion, it enters the shear-dominated regime. At this point, larger embryos take longer to double in mass than do smaller ones, although embryos of all masses continue their runaway growth relative to surrounding planetesimals; this phase of rapid accretion of planetary embryos is known as oligarchic growth (Kokubo & Ida 1998).

The self-limiting nature of runaway/oligarchic growth implies that massive planetary embryos form at regular intervals in semi-major axis. The agglomeration of these embryos into a small number of widely spaced terrestrial planets necessarily requires a stage characterized by large orbital eccentricities, significant radial mixing, and giant impacts. At the end of the runaway phase, most of the original mass is contained in the large bodies, so their random velocities are no longer strongly damped by energy equipartition with the smaller planetesimals. Mutual gravitational scattering can pump up the relative velocities of the planetary embryos to values comparable to the surface escape velocity of the largest embryos, which is sufficient to ensure their mutual accumulation into planets. The large velocities imply small collision cross-sections and hence long accretion times.

Once the planetary embryos have perturbed one another into crossing orbits, their subsequent orbital evolution is governed by close gravitational encounters and violent, highly inelastic collisions. This process has been studied using $N$-body integrations of planetary embryo orbits, which include the gravitational effects of the giant planets, but neglect the population of numerous small bodies which must also have been present in the terrestrial zone; physical collisions are assumed to always lead to accretion (i.e., fragmentation is not considered). Few bodies initially in the terrestrial planet zone are lost; in contrast, most planetary embryos in the asteroid region are ejected from the system by a combination of Jovian perturbations and mutual gravitational scatterings. As the simulations endeavor to reproduce our Solar System, they generally begin with about 2 $M_\oplus$ of material in the terrestrial planet zone, typically divided among several dozen or more protoplanets. The end result is the formation of two to five terrestrial planets on a timescale of about $1$–$2 \times 10^8$ years (Agnor, Canup, & Levison 1999; Chambers 2001). Some of these systems look quite similar to our Solar System, but most have fewer terrestrial planets which travel on more eccentric orbits. It is possible that the Solar System is, by chance, near the quiescent end of the distribution of terrestrial planets. Alternatively, processes such as fragmentation and gravitational interactions with a remaining population of small debris (or gas drag; Kominami & Ida 2004), thus far omitted from most calculations because of computational limitations, may lower the characteristic eccentricities and inclinations of the ensemble of terrestrial planets.

## 4. Terrestrial planet growth in binary star systems

More than half of all main sequence stars, and an even larger fraction of pre-main sequence stars, are in binary/multiple star systems (Duquennoy & Mayor 1991; Mathieu et al. 2000). At least 19 of the first 120 extrasolar planets to be detected are on so-called S-type orbits that encircle one component of a main-sequence binary star system (Eggenberger et al. 2004). The effect of the stellar companion on the formation of these planets, however, remains unclear. As discussed in Section 2, one planet has been confirmed in a P-type orbit which encircles the center of mass of PSR 1620–26, a radio pulsar binary comprised of a neutron star and a white dwarf in a ~191-day stellar orbit (Lyne et al.

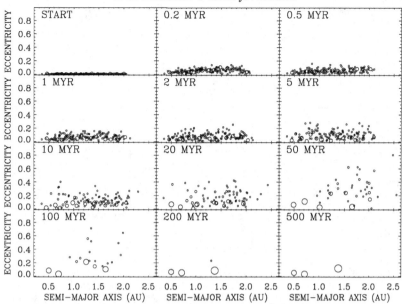

FIGURE 2. The temporal evolution of a circumstellar disk centered around $\alpha$ Centauri A, whose midplane was initially inclined by 15° to the stellar orbit, is shown (simulation A$i$15.3 in Quintana et al. 2002). The embryos' and planetesimals' eccentricities are displayed as a function of their semi-major axes, and the radius of each symbol is proportional to the radius of the body that it represents. After 200 Myr, four terrestrial planets had formed within 2 AU, accreting $\sim$88% of the initial disk mass. See Quintana (2004) for analogous diagrams of two additional systems with nearly identical initial conditions.

1988). Two substellar companions have been detected around the G6V star HD 202206, with minimum masses ($M \sin i$) of 17.4 $M_J$ at 0.83 AU and 2.44 $M_J$ at 2.55 AU (Udry et al. 2004). The inner companion is so massive that it is considered to be a brown dwarf, and it is likely that the outer companion formed from within a circumbinary (star-brown dwarf) disk (Correia et al. 2005). Planets have not been detected in P-type orbits around main-sequence binary stars, but short-period binaries are not included in precise Doppler radial-velocity search programs because of their complex and varying spectra.

The vast majority of theoretical effort to understand the formation of planetary systems has concentrated on isolated stars like our Sun. Binary star systems present a much higher dimensional phase space to cover (stellar mass ratio and orbital parameters), and orbital calculations in binary systems are more complicated than in systems that have one dominant mass. Nonetheless, zeroth-order questions about the dynamics of terrestrial planetary growth around single stars are now sufficiently well understood that it is worthwhile to investigate how the process of terrestrial planet growth differs in binary star systems.

The initial stages of terrestrial planet growth have been studied for gas-free circumprimary disks around 1 $M_\odot$ stars that have binary companions with masses in the range 0.1–1 $M_\odot$ by Whitmire et al. (1998) and for a gas-rich disk around $\alpha$ Centauri A by Marzari & Scholl (2000). Kortenkamp & Wetherill (2000) investigated this stage in the analogous problem of terrestrial planet growth within a gaseous disk containing pre-existing giant planets. Gas drag acts to reduce inclinations and align eccentricities of small solid bodies within binary star systems, thereby facilitating planetesimal formation. Planetesimals that are favorably located to survive this stage are likely to be less

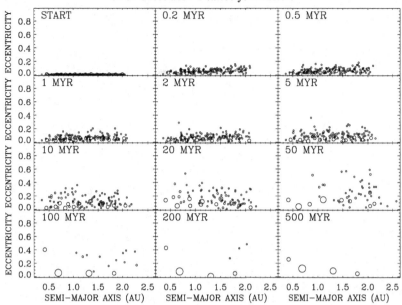

FIGURE 3. Same as Figure 2, except that the initial position of one of the planetesimals was moved by one meter along the direction of the orbit (simulation A$i$15_4 in Quintana et al. 2002). The differences between the simulations are produced by deterministic chaos, which implies that results of planetary accretion are extremely sensitive to changes in initial conditions. Thus, results are only valid in a statistical sense, and several simulations with very similar initial conditions must be run in order to adequately sample the distribution of possible outcomes. See Quintana (2004) for analogous diagrams of two additional systems with nearly identical initial conditions.

affected by the binary companion in the runaway and oligarchic growth stages, so they are probably able to grow into planetary embryos.

Quintana et al. (2002) performed dynamical simulations of the growth from planetary embryos into planets around each star within the $\alpha$ Centauri AB binary star system. These simulations begin with an initial disk composed of 140 planetesimals each of mass 0.00933 $M_\oplus$ and 14 planetary embryos each of mass 0.0933 $M_\oplus$. The initial masses and orbital parameters of the planetesimals and embryos were virtually identical to those used for the most 'successful' simulations of terrestrial planet growth in our Solar System (Chambers 2001). Planets with orbits similar to the giant planets within our Solar System could not be present in the $\alpha$ Centauri system. Nonetheless, if a planet formed near 1 AU, it could potentially survive for eons. Quintana et al. found that when the disk was inclined to the binary orbit by up to 30° (or when the initial inclination was equal to 180°), three to five terrestrial planets formed, and their configuration resembled that of the planets in our Solar System (Figs. 2 and 3). In contrast, terrestrial planet growth around a star lacking both stellar and giant planet companions is slower and extends to larger semi-major axis for the same initial disk parameters. When the disk was initially inclined to the binary orbit by 45° or 60°, a substantial fraction of the disk particles were lost to the inner star, and typically only one or two planets of significant size remained at the end of the calculation (Fig. 4). Complementary simulations of terrestrial planet growth around $\alpha$ Centauri A by Barbieri et al. (2002), who varied the initial distribution of planetesimals substantially, yielded results that are consistent with those of Quintana et al. A study of terrestrial planet growth around individual components

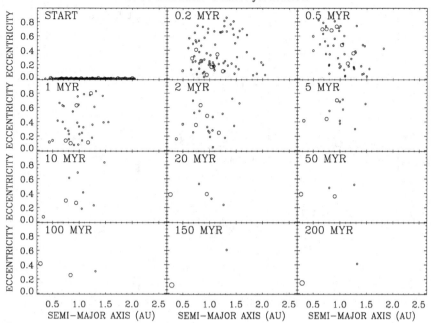

FIGURE 4. The temporal evolution of our standard planetesimal disk centered around $\alpha$ Cen A, with its midplane initially inclined at $i = 60°$ to the stellar orbit, is shown here (simulation A$i60l\_4$ in Quintana et al. 2002). The binary companion's high initial inclination causes large variations in the eccentricity of each planetesimal and embryo, and most of the mass is perturbed into the central star within the first few million years. An evolution plot for a simulation with nearly identical initial conditions that yields a smaller single planet is shown in Figure 6 of Quintana et al., and one for a simulation that ends up with a planet plus two planetesimals is shown by Lissauer et al. (2004).

in wide binary systems with stellar properties different from the $\alpha$ Centauri system is currently in progress; see Lissauer et al. (2004) for some preliminary results.

Quintana & Lissauer (2006) have performed an analogous investigation of the late stages of planetary growth in P-type orbits about close binary stars. Terrestrial planets similar to those formed in simulations of accretion around the Sun with giant planets perturbing the system can form around sufficiently tight binary stars. Increasing the initial binary eccentricity $e_B$ yields terrestrial planet systems that tend to be more sparce and have more orbitally diverse orbits. Binary stars with apastron distance $Q_B \equiv a_B(1 + e_B) \leqslant 0.2$ AU, where $a_B$ is the initial binary semi-major axis and the $e_B$ is initial binary eccentricity, do not, statistically, have very different effects on the planetary accretion disk under study (Fig. 5). The effect of the stellar perturbations on the planetesimal disk, however, become evident in simulations with $a_B = 0.2$ AU and $e_B = 0.5$, and in systems with $a_B > 0.2$ AU. From 1–2 terrestrial planets (more massive than the planet Mercury) formed in systems with $a_B > 0.2$ AU (Fig. 6), and more than half of the initial planetesimal/embryo disk mass was ejected from these systems.

## 5. Giant planet formation models

The observation that the mass function of young objects in star-forming regions extends down through the brown dwarf mass range to below the deuterium burning limit (Zapatero Osorio et al. 2000), together with the lack of any convincing theoretical reason to believe that the collapse process that leads to stars cannot also produce substellar

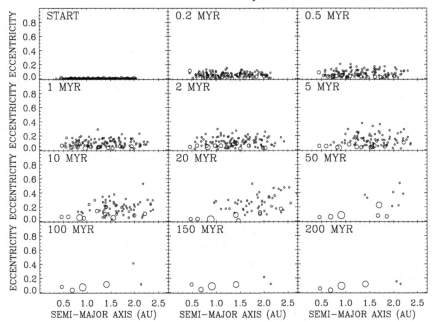

FIGURE 5. Our standard planetesimal disk placed around two 0.5 $M_\odot$ stars with $a_B = 0.2$ AU and $e_B = 0$ (run CB_.2_0_.5_c in Quintana & Lissauer 2006) results in the formation of a terrestrial planet system similar to that formed around a single 1 $M_\odot$ star.

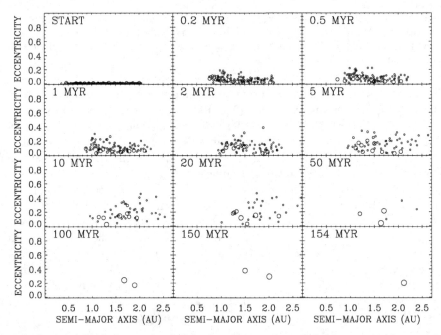

FIGURE 6. Our standard planetesimal disk placed around two 0.5 $M_\odot$ stars with $a_B = 0.2$ AU and $e_B = 0.5$ (run CB_.2_.5_.5_c in Quintana & Lissauer 2006) results in the ejection of much of the material in the terrestrial planet region, leaving less mass in the terrestrial planets than in simulations of growth from an analogous disk around a single 1 $M_\odot$ star. A simulation with very similar initial conditions that yields only a single terrestrial planet is shown by Lissauer et al. (2004).

objects (Wuchterl & Tscharnuter 2003), strongly implies that most isolated (or distant companion) brown dwarfs† and isolated high planetary mass objects formed via the same collapse process as do stars.

By similar reasoning, the 'brown dwarf desert,' the profound dip in the mass function of companions orbiting within several AU of Sun-like stars (Marcy et al. 2004), strongly suggests that the vast majority of extrasolar giant planets formed via a mechanism different from that of stars. Within our Solar System, bodies up to the mass of Earth consist almost entirely of condensable (under reasonable protoplanetary disk conditions) material, and the fraction of highly volatile gas increases with planet mass through Uranus/Neptune, to Saturn and finally Jupiter (which is still enriched in condensables at least threefold compared to the Sun), arguing for a unified formation scenario for all of the planets and smaller bodies. The continuum of observed extrasolar planetary properties, which stretches to systems not very dissimilar to our own, suggests that extrasolar planets formed as did the planets within our Solar System.

Models for the formation of gas giant planets were reviewed by Wuchterl, Guillot, & Lissauer (2000). Star-like direct quasi-spherical collapse is not considered viable, both because of the observed brown dwarf desert mentioned above and the theoretical arguments against the formation of Jupiter-mass objects via fragmentation (Bodenheimer et al. 2000a). The theory of giant planet formation that is favored by most researchers is the *core-nucleated accretion model*, in which the planet's initial phase of growth resembles that of a terrestrial planet, but the planet becomes sufficiently massive (several $M_\oplus$) that it is able to accumulate substantial amounts of gas from the surrounding protoplanetary disk. The only other hypothesis receiving significant attention is the *gas instability model*, in which a giant planet forms directly from the contraction of a clump that was produced via a gravitational instability in the protoplanetary disk.

Numerical calculations on gravitationally unstable disks by Adams & Benz (1992), recent work by Boss (2000), and Mayer et al. (2002) have revived interest in the gas instability model. However, these instabilities can only occur in disks with atypical physical properties (Ravikov 2005). Additionally, the gas instability hypothesis only accounts for massive stellar-composition planets, requiring a separate process to explain the smaller bodies in our Solar System and the heavy element enhancements in Jupiter and Saturn. It is particularly difficult to account for the existence of intermediate objects like Uranus and Neptune in such a scenario. See Durisen (this volume) for a more extensive discussion of the gas instability model.

The core nucleated accretion model relies on a combination of planetesimal accretion and gravitational accumulation of gas. In this theory, the core of the giant planet first forms by accretion of planetesimals, while only a small amount of gas is accreted. Core accretion rates depend upon the surface mass density of solids in the disk and physical assumptions regarding gas drag, planetary migration, etc. (Lissauer 1987; Pollack et al. 1996; Inaba et al. 2003; Alibert et al. 2005). The escape velocity from a planetary embryo with $M > 0.1\ M_\oplus$ is larger than the sound speed in the gaseous protoplanetary disk. Such a growing planetary core first attains a quasi-static atmosphere (that undergoes

---

† Following Lissauer (2004), these definitions are used throughout this chapter:
*Star*: self-sustaining fusion is sufficient for thermal pressure to balance gravity.
*Stellar remnant*: dead star—'no' more fusion (i.e., thermal pressure sustained against radiative losses by energy produced from fusion is no longer sufficient to balance gravitational contraction).
*Brown dwarf*: substellar object with substantial deuterium fusion (more than half of the object's original inventory of deuterium is ultimately destroyed by fusion).
*Planet*: negligible fusion ($< 13M_J$) + orbits star(s) or stellar remnant(s).

Kelvin-Helmholtz contraction as the energy released by the accretion of planetesimals) and gas is radiated away at the photosphere.

The contraction timescale is determined by the efficiency of radiative transfer, which is relatively low in some regions of the envelope. Spherically symmetric (1-D) models show that the minimum contraction timescale is a rapidly decreasing function of the core's mass (Pollack et al. 1996). The gas accretion rate, which is initially very slow, accelerates with time and becomes comparable to the planetesimal bombardment rate after the core has grown to $\sim 10$ $M_\oplus$. Once the gaseous component of the growing planet exceeds the solid component, gas accretion becomes very rapid, and leads to a runaway accumulation of gas.

The composition of the atmosphere of a giant planet is largely determined by how much heavy material was mixed with the lightweight material in the planet's envelopes. Once the core mass exceeds $\sim 0.01$ $M_\oplus$, the temperature becomes high enough for water to evaporate into the protoplanet's envelope. As the envelope becomes more massive, late-accreting planetesimals sublimate before they can reach the core, thereby enhancing the heavy element content of the envelope considerably.

The fact that Uranus and Neptune contain much less $H_2$ and He than Jupiter and Saturn suggests that Uranus and Neptune never quite reached runaway gas accretion conditions, possibly due to a slower accretion of planetesimals (Pollack et al. 1996). The rate at which accretion of solids takes place depends upon the surface density of condensates and the orbital frequency, both of which decrease with heliocentric distance. Alternatively/additionally, Uranus and Neptune may have avoided gas runaway as a result of the removal of gas from the outer regions of the disk via photoevaporation (Hollenbach et al. 2000). Additional theoretical difficulties for forming planets at Uranus/Neptune distances have been addressed by Lissauer et al. (1995) and Thommes et al. (2003). New models are being proposed to address these problems by allowing rapid runaway accretion of a very small number of planetary embryos beyond 10 AU. In the model presented by Weidenschilling (2005), an embryo is scattered from the Jupiter-Saturn region into a massive disk of small planetesimals. Goldreich et al. (2004) propose that planetesimals between growing embryos are ground down to very small sizes and are forced into low inclination, nearly circular orbits by frequent mutual collisions. Planetary embryos can accrete rapidly in such dynamically cold disks as those in the models of Weidenschilling and of Goldreich et al. Alternatively, Thommes et al. (2003) suggest that Uranus and Neptune accreted much closer to the Sun than these planets are at present, and were subsequently scattered out to their current locations by gravitational perturbations of Jupiter and Saturn (see also Tsiganis et al. 2005).

During the runaway planetesimal accretion epoch, the protoplanet's mass increases rapidly. The internal temperature and thermal pressure increase as well, preventing substantial amounts of nebular gas from falling onto the protoplanet. When the planetesimal accretion rate decreases, gas falls onto the protoplanet more rapidly. The protoplanet accumulates gas at a gradually increasing rate until its gas component is comparable to its heavy element mass. The key factor limiting gas accumulation during this phase of growth is the protoplanet's ability to radiate away energy and contract (Pollack et al. 1996; Hubickyj, Bodenheimer, & Lissauer 2005). The rate of gas accretion then accelerates more rapidly, and a gas runaway occurs. The gas runaway continues as long as there is gas in the vicinity of the protoplanet's orbit. The protoplanet may cut off its own supply of gas by gravitationally clearing a gap within the disk (Lin & Papaloizou 1979). Such gaps have been observed around small moons within Saturn's rings (Showalter 1991). D'Angelo, Kley, & Henning (2003) are using a 3-D adaptive mesh refinement code to follow the flow of gas onto an accreting giant planet. Models such as this will eventually allow the determination of final planetary mass as a function of the time-varying properties (density, temperature, viscosity, longevity, etc.) of the surrounding disk. Such a self-regulated growth limit provides a possible explanation to the observed mass distribution of extrasolar giant planets. Alternatively, the planet may accumulate all of the gas that remains in its region of the protoplanetary disk.

A major uncertainty associated with the emergence of planets is their predicted orbital migration as a consequence of the gravitational torque between the disk and the planet (Goldreich & Tremaine 1980; Ward 1986; Bate et al. 2003). Planetary orbits can migrate towards (or in some circumstances away from) their star as a consequence of angular momentum exchange between the protoplanetary disk and the planet. Planets that are more massive than Mars may be able to migrate substantial distances prior to the dispersal of the gaseous disk. Thus, it is quite possible that giant planets may form several AU from their star and then migrate inwards to the locations at which most extrasolar planets have been observed. Disk-induced migration is considered to be the most likely explanation for the 'giant vulcan' planets with periods of less than a week, because *in situ* formation of such objects is quite unlikely (Bodenheimer et al. 2000b). Livio & Pringle (2003) find no basis to suggest that planetary migration is sensitive to disk metallicity, and conclude that higher metallicity probably results in a higher likelihood of (giant) planet formation. The difficulty with the migration models is that they predict that planets should migrate *too rapidly*, especially in the Earth-to-Neptune mass range that planetary cores grow through in the core-nucleated accretion scenario. Moreover, because predicted migration rates increase as a planet moves inwards, most migrating planets should be consumed by their star. However, a planet may end up in very close 51 Peg-like orbits if stellar tides can counteract the migration, or if the disk has a large inner hole (Lin et al. 2000). Resolution of this rapid migration dilemma may require the complete and nonlinear analysis of the disk response to the protoplanet in the corotation regions. See Ward & Hahn (2000), Masset & Papaloizou (2003), and Thommes & Lissauer (2005) for more extensive discussions of planetary migration.

Many of the known extrasolar giant planets move on quite eccentric ($0.2 < e < 0.7$) orbits. These orbital eccentricities may be the result of stochastic gravitational scatterings among massive planets which have subsequently merged or been ejected to interstellar space (Weidenschilling & Marzari 1996; Levison, Lissauer, & Duncan 1998; Ford, Havlickova, & Rasio 2001), by perturbations of a binary companion (Holman, Touma, & Tremaine 1997), or by past stellar companions if the now-single stars were once members of unstable multiple-star systems (Laughlin & Adams 1998). However, as neither scattering nor migration offer a simple explanation for those planets with nearly circular orbits and periods from a few weeks to a few years, the possibility of giant planet formation quite close to stars should not be dismissed (Bodenheimer et al. 2000b).

Most of the observed extrasolar giant planets orbit between a few tenths of an AU and a few AU from their star, i.e., they are located much closer to their stars than Jupiter is from our Sun. These planets may have formed farther from their star and migrated inwards, but without a stopping mechanism, which isn't known at these distances, they would have fallen into the star. Lissauer (2001) suggested that the orbits could be explained if disks cleared from the inside outwards, leaving the planets stranded once they were too far interior to the disk for strong gravitational coupling to persist. Observations of the 2:1 resonant planets orbiting GJ 876 by Marcy et al. (2001; see also Rivera et al. 2005) support such a model, as do data which imply that the star CoKu Tau/4 has a disk with an inner hole (Forrest et al. 2004).

## 6. Conclusions: Summary of giant planet formation models

The smoothness of the distribution of masses of young M stars, free-floating brown dwarfs, and even free-floating objects somewhat below the deuterium-burning limit, argues strongly that these bodies formed in the same manner, i.e., via collapse, in some cases augmented by fragmentation. In contrast, the mass gap in nearby companions to Sun-like stars (the brown dwarf desert) is convincing evidence that (at least most of) the known giant planets formed in a different manner.

Various models for giant planet formation have been proposed. According to the prevailing core nucleated accretion model, giant planets begin their growth by the accumulation of small solid bodies, as do terrestrial planets. However, unlike terrestrial planets, the growing giant planet cores become massive enough that they are able to accumulate substantial amounts of gas before the protoplanetary disk dissipates. The primary question regarding the core-accretion model is whether planets with small cores can accrete very massive gaseous envelopes within the lifetimes of gaseous protoplanetary disks.

The main alternative giant planet formation scenario is the disk instability model, in which gaseous planets form directly via gravitational instabilities within protoplanetary disks. Formation of giant planets via gas instability has never been demonstrated for realistic disk conditions. Moreover, this model has difficulty explaining the supersolar abundances of heavy elements in Jupiter and Saturn, and it does not explain the origin of planets like Uranus and Neptune. Nonetheless, it is possible that some giant planets form via disk instability.

Most models for extrasolar giant planets suggest that they formed as Jupiter and Saturn are believed to have (in nearly circular orbits, far enough from the star that ice could condense), and subsequently migrated to their current positions, although some models suggest *in situ* formation. Gas giant planet formation may (or may not) be common, because the gas within most of protoplanetary disks could be depleted before solid planetary cores grow large enough to gravitationally trap substantial quantities of gas. Additionally, an unknown fraction of giant planets migrate into their star and are consumed, or are ejected into interstellar space via perturbations of neighboring giant planets—so even if giant planet formation is common, these planets may be scarce.

This work was supported by NASA's Solar System Origins Program under RTOP 188-07-21-03.

### REFERENCES

ADACHI, I., HAYASHI, C., & NAKAZAWA, K. 1976 *Prog. Theor. Phys.* **56**, 1756.

ADAMS, F. C. & BENZ, W. 1992. In *Complementary Approaches to Double and Multiple Star Research* (eds. H. A. McAlister & W. I. Hartkopf). ASP Conf. Ser. 32, p. 170. ASP.

AGNOR, C. B., CANUP, R. M., & LEVISON, H. F. 1999 *Icarus* **142**, 219.

ALENCAR, S. H. & BATALHA, C. 2002 *ApJ* **571**, 378.

ALIBERT, Y., MOUSIS, O., MORDASINI, C., & BENZ, W. 2005 *ApJ* **626**, L57.

ALONSO, R., ET AL. 2004 *ApJ* **613**, L153.

ANDRÉ, P., WARD-THOMPSON, D. & BARSONY, M. 2000. In *Protostars and Planets IV* (eds. V. Mannings, S. Russel, & A. P. Boss). p. 59. Univ. Arizona Press.

BARBIERI, M., ET AL. 2002 *A&A* **396**, 219.

BARNES, R. 2004. *Ph.D. Thesis.* University of Washington.

BARNES, R., QUINN, T. R., LISSAUER, J. J., & RICHARDSON, D. C. 2006 *ApJ*, in preparation.

BATE, M. R., LUBOW, S. H., OGILVIE, G. I., & MILLER, K. A. 2003 *MNRAS* **341**, 213.

BODENHEIMER, P., BURKET, A., KLEIN, R., & BOSS, A. P. 2000a. In *Protostars and Planets IV* (eds. V. Mannings, S. Russel, & A. P. Boss). p. 675. Univ. Arizona Press.

BODENHEIMER, P., HUBICKYJ, O., & LISSAUER, J. J. 2000b *Icarus* **143**, 2.

BOSS, A. P. 2000 *ApJ* **536**, L101.

BURROWS, A., SUDARSKY, D., & HUBBARD, W. B. 2003 *ApJ* **594**, 545.

CASSEN, P., SHU, F. H., & TEREBEY, S. 1985. In *Protostars and Planets II* (eds. D. C. Black & M. S. Matthews). p. 448. Univ. Arizona Press.

CHAMBERLAIN, T. C. 1905. In *Carnegie Institution Year Book 3 for 1904.* p. 195. Carnegie Inst. of Washington D.C.

CHAMBERS, J. E. 2001 *Icarus* **152**, 205.

CHAMBERS, J. E., QUINTANA, E. V., DUNCAN, M. J., & LISSAUER, J. J. 2002 *AJ* **123**, 2884.

CHARBONNEAU, D., BROWN, T. M., LATHAM, D. W., & MAYOR, M. 2000 *ApJ* **529**, L45.

CHIANG, E. & GOLDREICH, P. 1997 *ApJ* **490**, 368.

CORREIA, A. C. M., UDRY, S., MAYOR, M., LASKAR, J., NAEF, D., PEPE, F., QUELOZ, D., & SANTOS, N. C. 2005 *A&A*, **440**, 751.

CUZZI, J. N., HOGAN, R. C., PAQUE, J. M., & DOBROVOLSKIS, A. R. 2001 *ApJ* **546**, 496.

D'ANGELO, G., KLEY, W., & HENNING, T. 2003 *ApJ* **586**, 540.

DUQUENNOY, A. & MAYOR, M. 1991 *A&A* **248**, 485.

EGGENBERGER, A., ET AL. 2004. In *Extrasolar Planets: Today and Tomorrow* (eds. J. P. Beaulieu, A. L. Etangs, & C. Terquem) ASP Conf. Ser. 321, p. 93. ASP.

FISCHER, D. A. & VALENTI, J. 2005 *ApJ* **622**, 1102.

FORD, E. B., HAVLICKOVA, M., & RASIO, F. A. 2001 *Icarus* **150**, 303.

FORREST, W. J., ET AL. 2004 *ApJ* **154**, 443.

GILLILAND, R. L., ET AL. 2000 *ApJ* **545**, L47.

GOLDREICH, P. & TREMAINE, S. 1980 *ApJ* **241**, 425.

GOLDREICH, P. & TREMAINE, S. 2004 *A&A* **42**, 549.

GOLDREICH, P. & WARD, W. R. 1973 *ApJ* **183**, 1051.

GONZALEZ, G. 2003 *Revs. Mod. Phys.* **75**, 101.

GOODMAN, A. A., BENSON, P. J., FULLER, G. A. & MYERS, P. C. 1993 *ApJ* **406**, 528.

GOODMAN, J. & PINDOR, B. 2000 *Icarus* **148**, 537.

GREENBERG, R., WACKER, J. F., HARTMAN, W. K. & CHAPMAN, C. R. 1978 *Icarus* **35**, 1.

HAYASHI, C., NAKAZAWA, I., & ADACHI, I. 1977 *Publ. Astron. Soc. Jpn.* **29**, 163.

HEWINS, R., JONES, R., & SCOTT, E. (EDS.) 1996. *Chondrules and the Protoplanetary Disk.* Cambridge Univ. Press.

HOLLENBACH, D., YORKE, H. W. & JOHNSTONE, D. 2000. In *Protostars and Planets IV* (eds. V. Mannings, S. S. Russell, & A. P. Boss). p. 401. Univ. Arizona Press.

HOLMAN, M. J., TOUMA, J., & TREMAINE, S. 1997 *Nature* **386**, 254.

HUBBARD, W. B., PODOLAK, M., & STEVENSON, D. J. 1995. In *Neptune and Triton* (ed. D. P. Cruikshank). p. 109. Univ. Arizona Press.

HUBICKYJ, O., BODENHEIMER, P., & LISSAUER, J. J. 2005 *Icarus*, **179**, 415.

INABA, S., WETHERILL, G. W., & IKOMA, M. 2003 *Icarus* **166**, 46.

KOKUBO, E. & IDA, S. 1998 *Icarus* **131**, 171.

KOLVOORD, R. A. & GREENBERG, R. 1992 *Icarus* **98**, 2.

KOMINAMI, J. & IDA, S. 2004 *Icarus* **167**, 231.

KORTENKAMP, S. J. & WETHERILL, G. W. 2000 *Icarus* **143**, 60.

LADA, E. A., STROM, K. M., & MYERS, P. C. 1993. In *Protostars and Planets III* (eds. E. H. Levy & J. I. Lunine). p. 245. Univ. Arizona Press.

LASKAR, J. 2000 *Phys. Rev. Lett.* **84**, 3240.

LAUGHLIN, G. & ADAMS, F. C. 1998 *ApJ* **508**, L171.

LEVISON, H. F., LISSAUER, J. J., & DUNCAN, M. J. 1998 *AJ* **116**, 1998.

LIN, D. N. C. & PAPALOIZOU, J. 1979 *MNRAS* **186**, 799.

LIN, D. N. C., PAPALOIZOU, J. C. B., TERQUEM, C., BRYDEN, G., & IDA, S. 2000. In *Protostars and Planets IV* (eds. V. Manning, S. Russell, & A. P. Boss). p. 1111. Univ. Arizona Press.

LISSAUER, J. J. 1987 *Icarus* **69**, 249.

LISSAUER, J. J. 1993 *ARA&A* **31**, 129.

LISSAUER, J. J. 1995 *Icarus* **114**, 217.

LISSAUER, J. J. 2001 *Nature* **409**, 23.

LISSAUER, J. J. 2004. In *Extrasolar Planets: Today and Tomorrow* (eds. J. P. Beaulieu, A. L. Etangs, & C. Terquem). ASP Conf. 321, p. 271. ASP.

LISSAUER, J. J., QUINTANA, E. V., CHAMBERS, J. E., DUNCAN, M. J., & ADAMS, F. C. 2004. In *Gravitational Collapse: From Massive Stars to Planets* (eds. G. García-Segura, G. Tenorio-Tagle, J. Franco, & H. W. Yorke). *Rev. Mex. Astron. Astrof.* **22**, 99.

LIVIO, M. & PRINGLE, J. E. 2003 *MNRAS* **346**, L42.

LYNE, A. G., BIGGS, J. D., BRINKLOW, A., McKENNA, J., & ASHWORTH, M. 1988 *Nature* **332**, 45.

MAC LOW, M. & KLESSEN, R. 2004 *Revs. Mod. Phys.* **76**, 125.

MARCY, G. W., BUTLER, R. P., FISCHER, D. A., & VOGT, S. S. 2004. In *Extrasolar Planets: Today and Tomorrow* (eds. J. P. Beaulieu, A. L. Etangs, & C. Terquem). ASP Conf. 321, p. 3. ASP.

MARCY, G. W., BUTLER, R. P., FISCHER, D. A., VOGT, S. S., LISSAUER, J. J., & RIVERA, E. J. 2001 *ApJ* **556**, 296.

MARCY, G. W., ET AL. 2005 *Prog. Theor. Phys. Supp.* **158**, 24.

MARZARI, F. & SCHOLL, H. 2000 *ApJ* **543**, 196.

MASSET, F. S. & PAPALOIZOU, J. C. B. 2003 *ApJ* **588**, 494.

MATHIEU, R. D., GHEZ, A. M., JENSEN, E. L. N., & SIMON, M. 2000. In *Protostars and Planets IV* (eds. V. Mannings, S. Russell, & A. P. Boss). p. 703. Univ. Arizona Press.

MAYER, L., QUINN, T., WADSLEY, J., & STADEL, J. 2002 *Science* **298**, 1756.

McCAUGHREAN, M. J., STEPELFELDT, K. R., & CLOSE, L. M. 2000. In *Protostars and Planets IV* (eds. V. Mannings, S. Russell. & A. P. Boss). p. 485. Univ. Arizona Press.

NAKANO, T. 1987 *MNRAS* **224**, 107.

OHTSUKI, K., STEWART, G. R., & IDA, S. 2002 *Icarus* **155**, 436.

POLLACK, J. B., HUBICKYJ, O., BODENHEIMER, P., LISSAUER, J. J., PODOLAK, M., & GREENZWEIG, Y. 1996 *Icarus* **124**, 62.

QUINTANA, E. V. 2004. *Ph.D. Thesis*, University of Michigan, Ann Arbor.

QUINTANA, E. V. & LISSAUER, J. J. 2006 *Icarus* **185**, 1.

QUINTANA, E. V., LISSAUER, J. J., CHAMBERS, J. E., & DUNCAN, M. J. 2002 *ApJ* **576**, 982.

RAFIKOV, R. R. 2005 *ApJ* **621**, L69.

RIVERA, E. J., ET AL. 2005 *ApJ* **634**, 625.

SAFRONOV, V. 1969. *Evolution of the protoplanetary cloud and formation of the Earth and planets.* Nauka Press; 1972 English translation: NASA TTF-677.

SANTOS, N. C., ISRAELIAN, G., MAYOR, M., REBOLO, R., & UDRY, S. 2003 *A&A* **398**, 363.

SATO, B., ET AL. 2005 *ApJ*, **633**, 465.

SHOWALTER, M. R. 1991 *Nature* **351**, 709.

SIGURDSSON, S. 1993 *ApJ* **415**, L36.

SIGURDSSON, S., RICHER, H. B., HANSEN, B. M., STAIRS, I., & THORSETT, S. E. 2003 *Science* **301**, 193.

STROM, S. E., EDWARDS, S., & SKRUTSKIE, M. F. 1993. In *Protostars and Planets III* (eds. E. H. Levy & J. I. Lunine). p. 837. Univ. Arizona Press.

THOMMES, E. W., DUNCAN, M. J., & LEVISON, H. F. 2003 *Icarus* **161**, 431.

THOMMES, E. W. & LISSAUER, J. J. 2005. In *Astrophysics of Life* (eds. M. Livio, I. N. Reid, & W. B. Sparks). p. 41. Cambridge Univ. Press.

TSIGANIS, K., GOMES, R., MORBIDELLI, A., & LEVISON, H. F. 2005 *Nature* **435**, 459.

UDRY, S., MAYOR, M., NAEF, D., PEPE, F., QUELOZ, D., SANTOS, N., & BURNET, M. 2002 *A&A* **390**, 267.

WARD, W. R. 1986 *Icarus* **67**, 164.

WARD, W. R. 2000. In *Origin of the Earth and Moon* (eds. R. M. Canup & K. Righter). p. 75. Univ. Arizona Press.

WARD, W. R. & HAHN, J. 2000. In *Protostars and Planets IV* (eds. V. Manning, S. Russell, & A. P. Boss) p. 1135. Univ. Arizona Press.

WEIDENSCHILLING, S. J. 1977 *MNRAS* **180**, 57.

WEIDENSCHILLING, S. J. 2005 *Space Sci. Rev.* **116**, 53.

WEIDENSCHILLING, S. J. & CUZZI, J. N. 1993. In *Protostars and Planets III* (eds. E. H. Levy & J. I. Lunine) p. 1031. Univ. Arizona Press.

WEIDENSCHILLING, S. J. & MARZARI, F. 1996 *Nature* **384**, 619.

WETHERILL, G. W. 1990 *Ann. Rev. Earth Planet. Sci.* **18**, 205.

WETHERILL, G. W. 1994 *Astrophys. Space Sci.* **212**, 23.

WETHERILL, G. W. & STEWART, G. R. 1989 *Icarus* **77**, 330.

WHITMERE, D. P., MATESE, J. J., CRISWELL, L., & MIKKOLA, S. 1998 *Icarus* **132**, 196.

WUCHTERL, G., GUILLOT, T., & LISSAUER, J. J. 2000. In *Protostars and Planets IV* (eds. V. Mannings, S. Russell, & A. P. Boss). p. 1081. Univ. Arizona Press.

WUCHTERL, G. & TSCHARNUTER, W. M. 2003 *A&A* **398**, 1081.

YOUDIN, A. N. & SHU, F. H. 2002 *ApJ* **580**, 494.

YOUNG, R. E. 2003 *New Astron. Revs.* **47**, 1.

ZAPATERO OSORIO, M. R., BÉJAR, V. J. S., MARTÍN, E. L., REBOLO, R., BARRADO, Y., NAVASCUÉS, D., BAILER-JONES, C. A., & MUNDT, R. 2000 *Science* **290**, 103.

# Core accretion–gas capture model for gas giant planet formation

## By OLENKA HUBICKYJ

UCO/Lick Observatory, University of California, Santa Cruz, CA 94064 and
NASA Ames Research Center, MS 245-3, Moffett Field, CA 94035, USA

The core accretion–gas capture model is generally accepted as the standard formation model for gas giant planets. It proposes that a solid core grows via the accretion of planetesimals, and then captures a massive envelope from the solar nebula gas. Simulations have been successful in explaining many features of giant planets. This chapter will present an overview of the historical and scientific developments of the model, a description of the computer code based on the core accretion hypothesis with a summary of results of recent computer simulations, and the effect the observational achievement of finding extrasolar planets has had on the core accretion–gas capture model.

## 1. Introduction

A decade ago, theoreticians modeling the formation of planets had only a sample of nine objects with which to compare their computed results. Today we have over 150 planets and 13 extrasolar planet systems (Marcy et al. 2005) that challenge the formation models and the previously developed scenarios. These planets are believed to be gas giants, and with new detection techniques, the number of planets and the range in planet masses will be expanded. Discoveries of Neptune-like planets (e.g., Butler et al. 2004) and transit planets (e.g., Richardson et al. 2004) have been announced. These planets are diverse in their characteristics (Bodenheimer & Lin 2002), and planet scientists are working to learn and explain their formation mechanism.

There are two major models for the formation of gas giant planets: (1) the core accretion–gas capture model (which will be referred to as the CAGC or the core accretion model), and (2) the gas instability model (sometimes referred to as the GGPP model). The core accretion model (Safronov 1969; Perri & Cameron 1974; Mizuno et al. 1978; Mizuno 1980; Bodenheimer & Pollack 1986; Pollack et al. 1996; Bodenheimer et al. 2000) proposes that giant planets form in two stages: the formation of a massive solid core by coagulation of planetesimals in the solar nebula, followed by the gravitational capture by the core of a massive envelope from the solar nebula gas. The gas instability model is a single-stage model in which the solar nebula becomes gravitationally unstable and rapidly collapses to form a gravitationally bound subcondensation known as a giant gaseous protoplanet (Kuiper 1951; Cameron 1978; DeCampli & Cameron 1979; Boss 1998, 2000; Mayer et al. 2002, 2004; Pickett et al. 2003; Rice & Armitage 2003; Rice et al. 2003; Boss 2003).

The core accretion model is generally accepted as the more likely scenario of the two formation theories. In the past few years, computer simulations based on the CAGC model have been quite successful in explaining many features of the gas giant planets in the Solar System (Pollack et al. 1996; hereafter referred to as Paper 1) and *in situ* formation of companions to 51 Peg, $\rho$ CrB, and 47 UMa (Bodenheimer et al. 2000). In the past, the core accretion model had difficulties making planets in a short time and with small core masses; the smaller the core mass, the longer the formation time. However, recent calculations (Hubickyj et al. 2005; hereafter referred to as HBL05) demonstrate that models of Jupiter can be computed well within the observational limits and those set by interior models of Jupiter based on observations.

The evolution of a gas giant planet in the core accretion model is described in Bodenheimer et al. (2000) and is viewed to occur in the following sequence:

(1) Dust particles in the solar nebula form planetesimals that accrete, resulting in a solid core surrounded by a low-mass gaseous envelope. Solid runaway accretion occurs, during which the gas accretion rate is much slower than that of solids. As the solid material in the feeding zone is depleted, the solid accretion rate is reduced. The gas accretion steadily increases, and eventually becomes greater than the solid accretion rate.

(2) The protoplanet continues to grow as the gas accretes at a relatively constant rate. The mass of the solid core also increases, but at a slower rate. Eventually, the core and envelope masses become equal (called the crossover mass, $M_{cross}$).

(3) Runaway gas accretion occurs and the protoplanet grows at rapidly accelerating rate. The evolution of stages (1)–(3) is referred to as the *nebular* stage, because the outer boundary of the protoplanetary envelope is in contact with the solar nebula, and the density and temperature at this interface are given nebular values.

(4) The gas accretion rate reaches a limiting value defined by the rate at which the nebula can transport gas to the vicinity of the planet. After this point, the equilibrium region of the protoplanet contracts inside the effective accretion radius (defined in Bodenheimer et al. 2000), and gas accretes hydrodynamically onto this equilibrium region. This part of the evolution is considered to be the *transition* stage.

(5) Accretion is stopped by either the opening of a gap in the disk as a consequence of the tidal effect of the planet, or by dissipation of the nebula. Once accretion stops, the planet enters the *isolation* stage.

(6) The planet contracts and cools to the present state at constant mass.

This chapter describes the scenario for the formation of the gas giant planets. Observational constraints on the model are summarized in Section 2. A short overview of the development of the CAGC model is described in Section 3, and a description of the computer simulation of the CAGC model is in Section 4. Recent results of the computer simulations are reported in Section 5, and conclusions and summaries are presented in Section 6.

## 2. Observational requirements for planet formation models

Up to the time of the discovery of the companion to 51 Peg (Mayor & Queloz 1995; Marcy & Butler 1995) the major effort of theoretical studies was to explain the nature of Jupiter, Saturn, Uranus, and Neptune. A successful formation theory of giant planets at that time needed to explain the following general characteristics:

(1) The observed bulk composition characteristics of Jupiter, Saturn, Uranus and Neptune (e.g., Pollack & Bodenheimer 1989). Specifically:

- the similarity of the total heavy element contents of the four giants,
- the very massive $H_2$ and He envelopes of Jupiter and Saturn and much less massive (but not negligible) gaseous envelopes of Uranus and Neptune, and,
- the enhancement of metals over solar abundance in the atmospheres of all four giant planets;

(2) Giant planets need to form quickly. Observed dust disks around young stellar objects indicate disk ages of <10 Myr (Strom et al. 1993).

The explanation for the bulk-composition characteristics comes as a natural consequence of the core accretion scenario. The results of the computer simulations of Paper 1 demonstrated that the mass of the solid component of the giant planets agreed with observations of the gas giants in the Solar System, and that the solid component is relatively independent of the position of the planet in the solar nebula. It was also shown that the formation time of their nominal model (within the context of the assumptions made in their calculations) was within the time compatible with the solar nebula dispersal timescale.

In the decade since the discovery of the companion to 51 Peg, the catalog of extrasolar planets has greatly expanded, there are more observations of protoplanetary disks, and improved interior models of Jupiter and Saturn. To be considered successful, the current observational characteristics that need to be explained by a theoretical formation model are:

(1) The observed bulk composition characteristics of Jupiter, Saturn, Uranus, and Neptune, with emphasis on the enhancement of metals over solar abundance in the atmospheres of all four giant planets (e.g., Young 2003). The interior models of Jupiter (Saumon & Guillot 2004) indicate that the total solid mass ranges from 8–39 $M_\oplus$, of which 0–11 $M_\oplus$ is concentrated in the core. Saturn models indicate a total heavy element mass of 13–28 $M_\oplus$, with a core mass between 9–22 $M_\oplus$. Uranus and Neptune models indicate heavy element masses ranging from 10–15 $M_\oplus$ and a gaseous mass between 2–4 $M_\oplus$ (Pollack & Bodenheimer 1989);

(2) The upper limit to the formation timescale is still 10 Myrs, *but* from observations of dust disks around young stellar objects (Cassen & Woolum 1999; Haisch et al. 2001; Lada 2003; Chen & Kamp 2004; Metchev et al. 2004) indicate disk ages of <10 Myr with a preference for the time to be 3 to 5 Myrs.

(3) The extrasolar planets exhibit a wide range of eccentricities and semi-major axes. In a few cases, there are long-period, low-eccentricity planets whose orbits are comparable to that of Jupiter. In addition, there is the observed correlation of extrasolar planets forming around parent stars with high metallicity. Santos et al. (2004) observed that 25–30% of the stars with [Fe/H] above 0.3 have a planet, whereas less than 5% of the stars with solar metallicity have observed companions.

Overall, it should be noted that at the time of this conference, *A Decade of Extrasolar Planets Around Normal Stars*, the CAGC model can explain the bulk compositional properties of the gas giants in the Solar System, and computer simulations based on this model can form planets in a timely fashion—namely, on a timescale of less than 10 Myr and even between 3 and 5 Myrs. In addition, the CAGC model is beginning to address the correlation of more frequent planet formation around parent stars with high metallicity (Kornet et al. 2005) and the effect of migration on the formation timescale (Alibert et al. 2005). Further discussion of how the CAGC model addresses these observational constraints is in Section 5.

## 3. Development of the CAGC model

Planet formation theories and models have been proposed, debated, and refined for quite a few decades. An overall discussion of planet formation is presented by J. Lissauer in this book, and a brief historical overview of the CAGC model is offered in this section.

The earliest theories proposed that planets formed from mass thrown off the Sun after it had condensed into its current state (Descartes 1644; Kant 1755; Laplace 1796; W. Herschel 1811). A rudimentary version of an accretion theory was proposed by Buffon (1749), in which he considered a "building up" formation process rather than a condensing mechanism, proposing that a comet passing close to the Sun pulled matter off the Sun, which then accreted into planets. It was understood early on that the planets were formed from the parent star's environment.

Over the next century, the methods by which the Sun's thrown-off material evolved into the planets was debated by geologists and Darwinian evolutionists (e.g., Chamberlin 1899; Moulton 1905), as well as by physicists and mathematicians (e.g., Jeffreys 1917, 1918; Jeans 1919; Russell 1935). During this time, two models persisted to trade off as the prominent planet formation theory: the nebular hypothesis (e.g., Kuiper 1951; Cameron 1978), referred to today as the gas instability model, and the accretion model (Urey 1951; Perri & Cameron 1974). This period, and the scientific milestones relevant to planet formation, is provided in more detail by Brush (1990).

The earliest quantitative work was undertaken by Safronov during the 1960s. He developed a model based on the work of Shmidt (1944), who postulated that the Sun captured material (a "protoplanetary cloud") from interstellar space. Safronov (1969) created an analytical formulation for the accumulation of solid particles from the protoplanetary cloud into planets. The Safronov accretion model and the burgeoning capabilities of computers prompted extensive work on planet formation. Wetherill (1980) was one of the earliest researchers who adopted Safronov's planetary accretion model for a computer simulation of Earth and terrestrial planet formation.

In tandem to Safronov's work, a series of papers by Kusaka et al. (1970), Hayashi (1981), and Nakagawa et al. (1981) investigated the growth of solid particles in the solar disk. Mizuno et al. (1978) included the effect of the gaseous nebula on the buildup of planetesimals into planets, and then Mizuno (1980) extended that accretion model to the formation of Jupiter and Saturn. Their work was based on the computation of a series of protoplanetary models of increasing core mass, with a gaseous envelope in hydrostatic equilibrium that extends out to the protoplanet's tidal radius. Mizuno determined that there is a maximum core mass, called the "critical" core mass, $M_{crit}$, for which a static solution for the envelope with a core mass greater than $M_{crit}$ was not possible. This value was determined to be $M_{crit} \approx 10\ M_\oplus$. They also found that $M_{crit}$ was insensitive to the distance from the Sun. The success of the study by Mizuno and his collaborators at Kyoto University marked a clear advantage of the accretion model over the gaseous condensation model.

About a decade before the discovery of the companion to 51 Peg, Bodenheimer & Pollack (1986) computed the first evolutionary calculation of gas giant planets based on the core accretion model. These models were based on an adapted stellar-evolution code with constant accretion rates. They found that the critical core mass was most sensitive to the rate of planetesimal accretion, namely, that $M_{crit}$ decreased as the planetesimal accretion rate was reduced. They corroborated Mizuno's results that $M_{crit}$ was not dependent on solar nebula boundary conditions, but that $M_{crit}$ was less sensitive to micron-sized grains in the envelope than was determined in Mizuno's calculation.

Wuchterl (1991a,b) used a radiation-hydrodynamics code rather than a quasi-static one used by previous investigators to analyze the core accretion model. He found that once the envelope mass became comparable to the core mass, a dynamical instability develops that results in the ejection of much of the envelope.

Within the last two decades, the CAGC model has become a sophisticated model, with computer simulations that explain many features of the gas giant planets (e.g., Paper 1; Bodenheimer et al. 2000; HBL05). Interior model calculations of the gas giant planets (e.g., Hubbard et al. 1999; Saumon & Guillot 2004) are an important component to the general investigation of gas giant planet formation. Based on actual observations of the giant planets in the Solar System (e.g., gravitational moments), substantial information about the presence of a solid core—and the size and composition of it—can be extracted when structural model parameters are matched with observed values (Marley et al. 1999). Though most of the theoretical work has been based on the gas giants in our Solar System, the understanding derived from these models has been extended to the extrasolar planets.


## 4. The CAGC computer model

At the time of this conference, there are four groups that have computer models based on the CAGC formation of gas giant planets: the collaborators at NASA-Ames Research Center and University of California at Santa Cruz (referred to ARC/UCSC group); the group in Japan; the group in Bern, Switzerland; and G. Wuchterl. The first three groups use a similar technique based on a modified stellar structure evolution code, and Wuchterl uses a fully hydrodynamical computer code to model the evolving protoplanet. The discussion of the computer code technique will be concentrated on the one used by the ARC/UCSC group, and variations on this work by others will be noted and described.

The ARC/UCSC code consists of three main components:

(1) *The calculation of the rate of solid accretion onto the protoplanet* with an updated version of the classical theory of planetary growth (Safronov 1969) to calculate the rate of growth of the solid core. The gravitational enhancement factor, which is the ratio of the total effective accretion cross section to the geometric cross section, is an analytical expression that was derived to fit the data from the numerical calculations of Greenzweig & Lissauer (1992), consisting of a large number of three-body (Sun, protoplanet, and planetesimal) orbital interaction simulations.

(2) *The calculation of the interaction of the accreted planetesimals with the gas in the envelope* (Podolak et al. 1988), which determines whether the planetesimals reach the core, are dissolved in the envelope, or a combination of the two. Calculations of trajectories of planetesimals through the envelope result in the radius in the envelope at which the planetesimal is captured (required to compute the accretion rate of the planetesimals), and the energy deposition profile in the envelope (required for the structure computation).

(3) *The calculation of the gas accretion rate and evolution of the protoplanet*, under the assumption that the planet is spherical and that the standard equations of stellar structure apply. The conventional stellar structure equations of conservation of mass and energy, hydrostatic equilibrium, and radiative or convective energy transport are used. The energy generation rate is the result of the accretion of planetesimals and the quasi-static contraction of the envelope.

The following assumptions were applied in the computer simulation:

(1) The growing protoplanet is a lone embryo, which is surrounded by a disk consisting of planetesimals with the same mass and radius. There is an initially uniform surface density in the region of the protoplanet.

(2) The protoplanet's feeding zone is assumed to be an annulus extending to a radial distance of about four Hill-sphere radii on either side of its orbit (Kary & Lissauer 1994), which grows as the planet gains mass. Planetesimals are spread uniformly over the zone and do not migrate into or out of the feeding zone.

(3) The equation of state is nonideal and the tables used are based on the calculations of Saumon et al. (1995), interpolated to a near-protosolar composition of $X = 0.74$, $Y = 0.243$, $Z = 0.017$. The opacity tables are derived from the calculations of Pollack et al. (1985) and Alexander & Ferguson (1994).

(4) The capture criterion includes planetesimals that deposit 50% or more of their mass into the envelope during their trajectory.

(5) Once the mass and energy profiles in the envelope have been determined, the planetesimals are assumed to sink to the core, liberating additional energy in the process.

(6) In order to account for the depletion of the planetesimal disk by accretion onto neighboring embryos, the rate of planetesimal accretion near gas runaway is limited to its value at crossover.

The inner and outer boundary conditions are set at the bottom and top of the envelope, respectively. The core is assumed to have a uniform density and to be composed of a combination of ice, CHON, and rock, depending on the conditions in the nebula in which the planet forms. The outer boundary condition of the protoplanet is applied in three ways, depending on the evolutionary stage of the planet (as noted in Section 10.1 and described in full in BHL00).

The calculations start at $t = 10^4$ yr with a core mass of 0.1 $M_\oplus$ and an envelope mass of $10^{-9}$ $M_\oplus$. These initial values were chosen for computational convenience, and the final results are insensitive to initial conditions. Near the end of the evolution of the protoplanet, the gas accretion rate is limited by the ability of the solar nebula to supply gas at the required rate. When this limiting rate is reached, the planet contracts inside its accretion radius (evolution enters the transition stage). The supply of gas to the planet is eventually assumed to be exhausted as a result of tidal truncation of the nebula, the removal of the gas by effects of the star, and/or the accretion of all nearby gas by the planet. The planet's mass levels off to the limiting value defined by the object that is being modeled. Since this process is not yet modeled in the ARC/UCSC code, the gas accretion rate onto the planet is assumed to reduce smoothly to zero as the limiting mass value is approached. The planet then evolves through the *isolated* stage, during which it remains at constant mass.

Figure 1 illustrates the typical nature of the simulations. The mass, luminosity, solid and gas accretion rates, and radii are plotted as a function of time. The solid core is accreted during Phase 1. This phase ends when the feeding zone is depleted of solids, thus the protoplanet reaches its isolation mass. Phase 2 is characterized by a steady rate of both solid and gas accretion, but with the gas rate being slightly greater. The duration of Phase 2 ends at the *crossover* point, when the gas mass is equal to the solid mass. It is evident that Phase 2 determines the overall timescale for the protoplanet to form, since this time is much longer than the times for the solid core to accrete and for the gas runaway to occur. Phase 3 is characterized by the gas runaway, which lasts until both the gas and solid accretion rates turn off. The protoplanet then cools and contracts to its presently observed state.

The Bern group's CAGC code (Alibert et al. 2005) is similar to the ARC/UCSC code *except* for the inclusion of the components that compute the evolution of the proto-planetary disk and the migration of the growing protoplanet. The disk structure and its evolution are based on the method of Papaloizou & Terquem (1999), which is in the framework of the $\alpha$-disk formulation of Shakura & Sunyaev (1973). Migration occurs when there is a dynamical tidal interaction of the growing protoplanet with the disk, which leads to two phenomena: inward migration and gap formation (Lin & Papaloizou 1979; Ward 1997; Tanaka et al. 2002). For low-mass planets, the tidal interaction is a linear function of mass and the migration is Type I (i.e., inward migration with no gap opening). Higher-mass planets open a gap, leading to a reduction of the inward migration; this is referred to as type II migration. The rest of the procedure used by the Bern group is similar to that used by the ARC/UCSC group: the solid accretion rate uses the gravitational enhancement factor based on that of Greenzweig & Lissauer (1992); the interaction between the infalling planetesimal and the atmosphere of the growing planet is based on the work of Podolak et al. (1988); and, the standard planetary structure

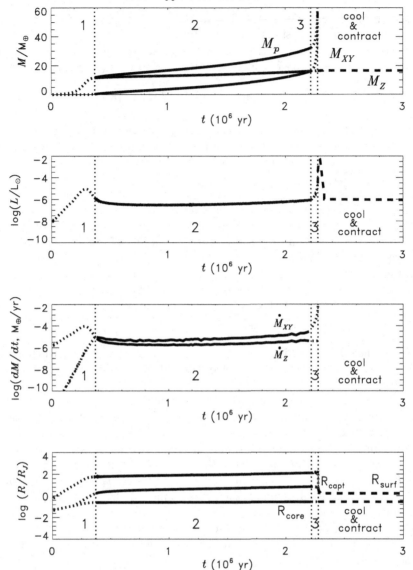

FIGURE 1. The mass (units of $M_\oplus$), the luminosity (units of $L_\odot$), the accretion rates (units of $M_\oplus$/year), and the radii (units of $R_{\rm Jup}$) are plotted as a function of time (units of million years) for the baseline case $10L^\infty$. The three phases of evolution and the cooling and contraction of the protoplanet are marked. *Dotted line*: Phase 1, *solid line*: Phase 2, *second dotted line*: Phase 3, *dashed line*: contraction and cooling.

and evolution scheme is applied. The rate for type I migration is a free parameter in the Alibert et al. (2005) calculations, therefore, the starting location of the embryo is adjusted for each choice of the migration rate, in order for the protoplanet to reach the crossover mass at 5.5 AU. The profile of the initial disk surface density, $\sigma_{\rm init,Z}$, is the same for each of the migration rates, namely a power law $\sigma_{\rm init,Z} \propto r^{-2}$, which was chosen to correspond to a case with $\sigma_{\rm init,Z} = 7.5$ g cm$^2$), which is about the twice the density of a minimum mass solar density (Paper 1).

The analysis of the gas flow in the envelope that was undertaken by Wuchterl (1991a,b, 1993) using a hydrodynamic code to study the flow velocity of the gas by solving an equation of motion for the envelope gas in the framework of convective radiation–fluid dynamics. This allows the study of the collapse of the envelope, of the accretion with finite Mach number, and of linear, adiabatic and nonlinear, non-adiabatic pulsational stability of the envelope. Furthermore, the treatment of convective energy transfer has been improved by calculations using a time dependent mixing-length theory of convection in hydrodynamics (Wuchterl 1995, 1999). Wuchterl starts his calculations at critical mass, $M_{\text{crit}}$, and finds that instead of collapsing as Mizuno surmised from his models, these envelopes begin to pulsate, resulting in mass loss. A state of quasi-equilibrium is achieved after the protoplanet loses a large fraction of its envelope mass. A major result of the hydrodynamical studies is that the protoplanet may pulsate and develop pulsation-driven mass loss. This leaves a planet with a low-mass envelope and properties similar to Uranus and Neptune, but the model does not account for Jupiter and Saturn. His results are contrary to the linear stability analysis of Tajima & Nakagawa (1997), who find the envelope models in Bodenheimer & Pollack (1986) to be dynamically stable. This issue with the quasi-static models is still unresolved.

The group in Japan continues to work on all aspects of planet formation. Kokubo & Ida (1998, 2002) studied the formation of planets and planet cores by oligarchic growth. Ikoma et al. (2000, 2001) examined the effect of the opacity of the grains in the protoplanet's growing envelope and the rate of solid accretion on formation timescales. Ida & Lin (2004) analytically investigated the migration of planets.

## 5. Recent results

There are two major computer simulation studies based on the CAGC model. The ARC/UCSC group examines the issue of the conditions for which Jupiter could have formed with a low mass core and a short formation timescale (HBL05). The Bern group considered the effects on giant planet formation with the inclusion of migration and protoplanetary evolution in the core accretion formation model (Alibert et al. 2005). Highlights of these two studies are presented below.

The ARC/UCSC group's simulations of the growth of Jupiter were computed for three parameters shown in Paper 1 to affect planet formation. The opacity produced by grains in the protoplanet's atmosphere was varied, and two different values (10 g cm$^2$ and 6 g cm$^2$) were used for the initial planetesimal surface density in the solar nebula. Additionally, halting the solid accretion at selected core-mass values during the protoplanet's growth was studied. Decreasing the atmospheric opacity due to grains emulates the settling and coagulation of grains within the protoplanetary atmosphere, and halting the solid accretion simulates the presence of a competing embryo. The effects of these parameters were examined in order to determine whether gas runaway can still occur for small-mass cores on a reasonable timescale (Ikoma et al. 2000).

The nomenclature for the simulations that denotes the parameters used in the computations is in the following form: $\sigma$-*opacity-cut*, where $\sigma$ is the initial surface density of planetesimals in the solar nebula with values 10 or 6 g cm$^2$; *opacity* is denoted by either $L$ for grain opacity at 2% of the interstellar value, $H$ for the full interstellar value, or $V$ for a variable (temperature dependent: $T < 350$ K ramping up to the full interstellar value for $T > 500$ K) grain opacity; and *cut* specifies the core mass (in units of $M_\oplus$), at which the planetesimal accretion rate is turned off. For cases with no solid accretion cutoff, *cut* is set to $\infty$. As an example, the model labelled $10L^\infty$ signifies that the simulation was

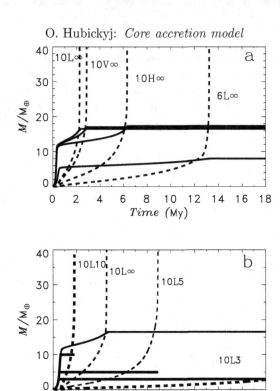

FIGURE 2. a) The masses (units of $M_\oplus$) of the four basic cases are plotted as a function of time (units of million years). The solid line denotes the mass of solids and the dashed line denotes the mass of gas. b) The masses of the baseline case $10L^\infty$ and the associated cutoff cases are plotted as a function of time. Units and line designations are the same as in a).

computed with $\sigma_{init,Z} = 10$ g cm$^2$, the grain opacity is 2% of the interstellar value, and there was no solid accretion cutoff.

Four series of simulations have been computed in this latest study. Each series consists of a run computed through the cooling and contracting of the protoplanet (i.e., Fig. 1), plus up to three runs with a cutoff of planetesimal accretion at a particular core mass. For these four basic cases, mass as a function of time is plotted in Figure 2a. The reduced grain opacities produce formation times that are less than half of that for models computed with full interstellar grain opacity values (see curves labelled $10H^\infty$ nd $10L^\infty$ in Fig. 2a). These models illustrate that the time spent in Phase 2 is decreased by $\sim$60% for models with the grain opacity set to 2% of the interstellar value. Therefore, another model was computed to determine if there was a temperature range for which the grain opacity had the most influence on the evolution time. This model ($10V^\infty$) was computed with the grain opacity set to 2% of the interstellar value for temperatures $\leq 350$ K and to the full interstellar value for temperatures $>500$ K, with interpolation in the intermediate region. The result of this calculation shows there is little difference from the model computed with the 2% interstellar value for the full temperature range. The reduction of opacity due to grains in the upper portion of the envelope with $T \leq 500$ K has the largest effect on the lowering of the formation time (see curves labelled $10H^\infty$, $10L^\infty$, and $10V^\infty$ in Fig. 2a). This result is profoundly important, especially with reference to the grain settling work of Podolak (2003), who has developed a numerical model for the growth and sedimentation of grains in a protoplanetary atmosphere, coupled with a procedure for calculating the opacity at each depth. These simulations indicate grain opacity values that are lower than the 2% ISM grain values we used in our most recent calculations.

Motivated by the interior models of Jupiter and Saturn by Guillot et al. (1997) and Saumon & Guillot (2004), who call for low solid mass cores for Jupiter and Saturn, the ARC/UCSC group studied the effect of the surface density of planetesimals and the effect of halting solid accretion on the formation of the protoplanet. Decreasing the surface density of planetesimals lowers the final core mass of the protoplanet, but increases the formation timescale considerably (see curves labelled $10L^\infty$ and $6L^\infty$ in Fig. 2a).

The effect of halting solid accretion is illustrated in Figure 2b. The plot shows the mass as a function of time for the baseline case $10L^\infty$ and the three associated runs for which the solid planetesimal accretion is turned off at core masses 10, 5, and 3 $M_\oplus$. It is clearly demonstrated that the time needed for a protoplanet to evolve to the stage of runaway gas accretion is reduced, provided the cutoff mass is sufficiently large. The overall results indicate that, with reasonable parameters and with the assumptions in the ARC/UCSC CAGC model code, it is possible that Jupiter formed via the core accretion process in 3 Myr or less.

Migration is more than likely a viable aspect of gas giant planet formation in explaining the wide range of eccentricities and semi-major axes, like the hot-Jupiter type extrasolar planets (Jupiter-sized planets found in orbits very close to their central star), deduced from the observations of extrasolar planets. Alibert et al. (2005) incorporated migration into their core accretion computer simulation and examined the effects on giant planet formation. Their results show that the formation timescale of gas giants is much shorter, by a factor of 10, when migration is included compared to *in situ* formation. A migrating embryo starting at 8 AU will migrate to 5.5 AU and reach crossover mass in ∼1 Myr, whereas the same embryo at 5.5 AU without migration and without disk evolution (i.e., *in situ* formation) reaches crossover mass in ∼30 Myr, a factor of 10 greater. The reason for this speed-up due to migration is quite simple. In CAGC formation models, the long formation timescale depends on the presence of Phase 2 occurring after the core is isolated at the end of Phase 1. A migrating embryo will never suffer isolation and will go directly from Phase 1 to Phase 3, reducing the formation time.

It should be noted that the *in situ* model computed by Alibert et al. (2005) should not be compared with the newer models in HBL05. In fact, the parameters chosen for the model discussed in Alibert et al. (2005) are those of a protoplanet growing in a protoplanetary disk which is twice that of a minimum-mass solar nebula—not three to four times as dense, as was used in the ARC/UCSC models.

Though most of the computational work is based on Jupiter and Saturn models, the understanding derived from these computer simulations can be applied to extrasolar planets—as evidenced by the observation trait of high-metal planets. It is not unreasonable to apply the conclusions to extrasolar planets. The observational studies of extrasolar planets have shown that planets are discovered much more frequently around metal-rich stars (Gonzalez 1998; Santos et al. 2001, 2004; Fischer & Valenti 2003, 2005). While this trend has been found to be consistent with simplified core-accretion models (Ida & Lin 2004; Kornet et al. 2005), it has also been suggested that the correlation is a result of preferential migration of planets in high-metallicity disks into the period range where they are observed (Sigurdsson et al. 2003). Sozzetti (2004) suggests that there is a correlation between observed orbital period and the host star metallicity in the sense that the higher-metallicity stars are more likely to have short-period planets. This tendency would be consistent with the migration scenerio; however, the correlation is weak, and Santos et al. (2003) do not find it. On the theoretical side, the simple model of Livio & Pringle (2003) results in only a small difference in migration rates in metal-rich and metal-poor disks, not sufficient to explain the trend seen in Fischer & Valenti (2003). Thus, this correlation is more likely to be a result of the formation mechanism itself. Although a higher-metallicity planet has a higher opacity in the envelope that results in longer formation times, increases in opacity by only a factor of two have a very small effect on the time; factors of 50 or more in opacity changes are required to make significant differences in formation times. More study on this topic is necessary.

## 6. Summary

Since Safronov's introduction of his planetesimal accretion model for terrestrial planets and Mizuno's extension to the formation of Jupiter and Saturn, theoreticians have made substantial progress in understanding planet formation in the last few decades. At the time of this conference, the CAGC simulations have provided the following conclusions that can be summarized as follows (Fig. 3):

(1) The opacity due to grains in the protoplanetary envelope has a major effect on the formation timescale, but no effect on the core mass. It is *not* possible for a gas giant with a small solid core to form on a short timescale for models computed with the grain opacity equal to that of typical interstellar material. For models computed with the grain opacity below interstellar values, formation times are short, but the final core mass is unaffected. The baseline case computed in HBL05 ($10L^\infty$) shows that Jupiter can be formed at 5 AU in just over 2 Myr, but the core mass is 16 $M_\oplus$.

(2) Halting the planetesimal accretion provides for formation times to be in the range of 1–4.5 Myr and for a core mass consistent with that of Jupiter, if the initial solid surface density in the disk is three times that of the minimum mass solar nebula.

(3) By reducing the initial solid surface density in the disk to two times that of the minimum-mass solar nebula, it is still possible to form Jupiter in less than 5 Myr if the core accretion is cutoff at 5 $M_\oplus$.

(4) All models satisfy the constraint that the total heavy element abundance is less than, or comparable to, the value deduced from observations of Jupiter. A few models have low heavy element abundance (3–5 $M_\oplus$), but it is quite reasonable to expect the planet to accrete more solids during or after rapid gas accretion, which is not taken into account in these models.

(5) The results of Paper 1 show that Phase 2, the early gas accretion phase before crossover mass is reached, essentially determines the timescale for the formation of a giant planet. However, the results presented in HBL05 indicate that for low atmospheric opacity and/or a cutoff in accretion of solids, Phase 2 can be relatively short. Thus, the time for Phase 1, the solid core accretion phase, may be the determining factor.

(6) Migration, which prevents the depletion of the feeding zone that occurs in *in situ* formation, appears to have a very important effect on the formation timescale by decreasing it by about a factor of 10 (Alibert et al. 2005), without having to consider massive disks (Lissauer 1987).

Although some old problems relating to planet formation have been resolved, there are others that still need investigating. Taking into account multiple embryos and a number of other physical processes, recent simulations of the core accretion process indicate that the core formation times are longer than those computed by the ARC/UCSC studies (Inaba et al. 2003; Thommes et al. 2003; Kokubo & Ida 2002). Thus, the question remains as to how large an enhancement of solid surface density, as compared to that in the minimum mass solar nebula, is needed to form a giant planet in a few Myr. Another problem is related to the migration of a gas giant planet. According to Type I migration calculations by Tanaka et al. (2002), it seems difficult to form a planet and prevent it from spiraling

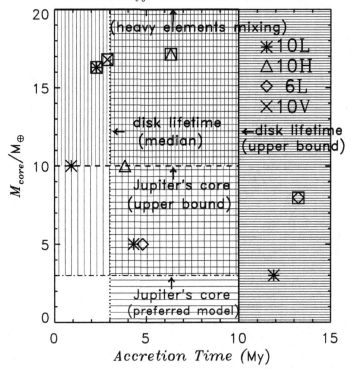

FIGURE 3. The core mass as a function of the time at the onset of runaway gas accretion for the four series of computed models. *Asterisks* denote models computed with $\sigma_{\mathrm{init},Z} = 10$ g cm$^2$ and grain opacity values that are 2% interstellar. *Triangles* denote models computed with $\sigma_{\mathrm{init},Z} = 10$ g cm$^2$ and full interstellar grain opacity. *Diamonds* denote models computed with $\sigma_{\mathrm{init},Z} = 6$ g cm$^2$ and 2% interstellar grain opacity. *X symbols* denote models computed with $\sigma_{\mathrm{init},Z} = 10$ g cm$^2$ and grain opacity at 2% of interstellar value for $T \leq 350$ K and full interstellar at $T > 500$ K. One model (10H5 with $M_{\mathrm{core}} = 5.$ M$_J$ and T$_{\mathrm{acc}} = 95.25$ Myr) is not plotted because it is outside the scope of the plot's limit. The *boxed symbols* denote the models for which there is *no* solid accretion cutoff; the mass at which the other runs are cut off is given by their vertical coordinate on the plot. The *solid vertical line* represents the approximate maximum observed disk lifetime around young stars. The *dotted line* represents the median observed disk lifetime. The *dot-dash* and the *dashed lines* represent, respectively, the preferred and the maximum value of Jupiter core mass, obtained from comparison of theoretical models with observations. The *bold* horizontal line at 20 $M_\oplus$, labelled "heavy elements mixing," represents a possible primordial core mass for Jupiter which was later eroded by convective mixing of heavy elements into the gaseous envelope. Regions above and to the right of the bold solid lines are strongly excluded by observations. Those failing less certain observational limits appear as partially shaded.

into its central star. Finally, problems that need to be investigated by models in the future include the formation of Uranus and Neptune, whose *in situ* core formation times seem to be too long for any reasonable disk model.

The author would like to thank the organizing committee for the privilege of attending this Workshop and being able to present these results. The author would also like to thank P. Bodenheimer and J. Lissauer for their collaboration on this CAGC model project. This work is supported by NASA grant NAG 5-9661 and NAG 5-13285 from the Origins of the Solar Systems Program.

# REFERENCES

ALEXANDER, D. R. & FERGUSON, J. W. 1994 *ApJ* **437**, 879.

ALIBERT, Y., MORDASINI, C., BENZ, W., & WINISDOERFFER, C. 2005 *A&A* **430**, 1133.

BODENHEIMER, P., HUBICKYJ, O., & LISSAUER, J. J. 2000 *Icarus* **143**, 2.

BODENHEIMER, P. & LIN, D. N. C. 2002 *Annu. Rev. Earth Planet. Sci.* **30**, 113.

BODENHEIMER, P. & POLLACK, J. B. 1986 *Icarus* **67**, 391.

BOSS, A. P. 1998 *ApJ* **503**, 923.

BOSS, A. P. 2000 *ApJ* **536**, L101.

BOSS, A. P. 2003 *ApJ* **599**, 577.

BRUSH, S. G. 1990 *Rev. Modn. Phys.* **62**, 43.

BUFFON, G.-L. LECLERC, COMTE DE 1749 *Histoire Naturelle, Générale et Particulière, avec la Description du Cabinet du Roi.* Imprimerie Royal.

BUTLER, R. P., VOGT, S. S., MARCY, G. W., FISCHER, D. A., WRIGHT, J. T., HENRY, G. W., LAUGHLIN, G., & LISSAUER, J. J. 2004 *ApJ* **617**, 580.

CAMERON, A. G. W. 1978 *Moon and Planets* **18**, 5.

CASSEN, P. & WOOLUM, D. 1999. In *Encyclopedia of the Solar System* (eds. P. R. Weissman, L. A. McFadden, & T. V. Johnson). p. 35. Academic Press.

CHAMBERLIN, T. C. 1899 *Science* **9**, 889; **10**, 11.

CHEN, C. H. & KAMP, I. 2004 *ApJ* **602**, 985.

DECAMPLI, W. M. & CAMERON, A. G. W. 1979 *Icarus* **38**, 367.

DESCARTES, R. 1644 *Principia Philosophiae.* Elzevir.

FISCHER, D. A. & VALENTI, J. A. 2003. In *Scientific Frontiers in Research on Extrasolar Planets* (eds. D. Deming & S. Seager). ASP Conf. Ser., Vol. 294, p. 117. ASP.

FISCHER, D. A. & VALENTI, J. A. 2005 *ApJ* **622**, 1102.

GONZALEZ, G. 1998 *A&A* **334**, 221.

GREENZWEIG, Y. & LISSAUER, J. J. 1992 *Icarus* **100**, 440.

GUILLOT, T., GAUTIER, D., & HUBBARD, W. B. 1997 *Icarus* **130**, 534.

HAISCH, K. E. JR., LADA, E. A., & LADA, C. J. 2001 *ApJ* **553**, L153.

HAYASHI, C. 1981 *Prog. Theor. Phys. Suppl.* **70**, 35.

HERSCHEL, W. 1811 *Philis. Trans. R. Soc. London* **101**, 269.

HUBBARD, W. B., GUILLOT, T., MARLEY, M. S., BURROWS, A., JUNINE, J. I., & SAUMON, D. 1999 *Plan. Space Sci.* **47**, 1175.

HUBICKYJ, O., BODENHEIMER, P., & LISSAUER, J. J. 2005 *Icarus* **179**, 415.

IDA, S. & LIN, D. N. C. 2004 *ApJ* **604**, 388.

IKOMA, M., EMORI, H., & NAKAZAWA, K. 2001 *ApJ* **553**, 999.

IKOMA, M., NAKAZAWA, K., & EMORI, H. 2000 *ApJ* **537**, 1013.

INABA, S., WETHERILL, G. W., & IKOMA, M. 2003 *Icarus* **166**, 46.

JEANS, J. H. 1919. *Problems of Cosmogony and Stellar Dynamics*, Adams Prize Essay for 1917. Cambridge University Press.

JEFFREYS, H. 1917 *Mem. R. Astron. Soc.* **62**, 1.

JEFFREYS, H. 1918 *MNRAS* **78**, 424.

KANT, I. 1755 *Allgemeine Naturgeschichte und Theorie des Himmels.* Johann Friederich Petersen.

KARY, D. M. & LISSAUER, J. J. 1994. In *Numerical Simulations in Astrophysics* (eds. J. Franco, S. Lizano, L. Aguilar, & E. Daltabuit. p. 364. Cambridge Univ. Press.

KOKUBO, E. & IDA, S. 1998 *Icarus* **131**, 171.

KOKUBO, E. & IDA, S. 2002 *ApJ* **581**, 666.

KORNET, K., BODENHEIMER, P., RÓŻYCZKA, M., & STEPINSKI, T. F. 2005 *A&A* **430**, 1133.

KUIPER, G. P. 1951. In *Astrophysics* (ed. J. A. Hynek). p. 357. McGraw-Hill.

KUSAKA, T., NAKANO, T., & HAYASHI, C. 1970 *Prog. Theor. Phys. Suppl.* **44**, 1580.

LADA, E. A. 2003 *BAAS* **35**, #24.06, 730.

LAPLACE, P. S. 1796 *Exposition de Système du Monde* Circle-Sociale; English translation by H. H. Harte in 1830: *The System of the World.* University Press.

LIN, D. N. C. & PAPALOIZOU, J. 1979 *MNRAS* **188**, 191.

LISSAUER, J. J. 1987 *Icarus* **69**, 249.

LIVIO, M. & PRINGLE, J. E. 2003 *MNRAS* **346**, L42.

MARCY, G. W. & BUTLER, R. P. 1995 *IAU Circ.* **6251**, 1.

MARCY, G. W., BUTLER, R. P., FISCHER, VOGT, S., WRIGHT, J. T., TINNEY, C. G., & JONES, H. R. A. 2005. In the proceedings of the Ringberg Workshop on Planet Formation, *Progress of Theoretical Physics Supplement*, No. 158, p. 24.

MARLEY, M. S. 1999. In *Encyclopedia of the Solar System* (eds. P. Weissman, L. A. McFadden, & T. V. Johnson). p. 339. Academic Press.

MAYER, L., QUINN, T., WADSLEY, J., & STADEL, J. 2002 *Science* **298**, 1756.

MAYER, L., QUINN, T., WADSLEY, J., & STADEL, J. 2004 *ApJ* **609**, 1045.

MAYOR, M. & QUELOZ, D. 1995 *Nature* **378**, 355.

METCHEV, S. A., HILLENBRAND, L., & MEYER, M. 2004 *ApJ* **600**, 435.

MIZUNO, H. 1980 *Prog. Theor. Phys.* **64**, 544.

MIZUNO, H., NAKAZAWA, K., & HAYASHI, C. 1978 *Prog. Theor. Phys.* **60**, 699.

MOULTON, F. R. 1905 *ApJ* **22**, 165.

NAKAGAWA, Y., NAKAZAWA, K., & HAYASHI, C. 1981 *Icarus* **45**, 517.

PAPALOIZOU, J. C. B. & TERQUEM, C. 1999 *ApJ* **521**, 823.

PERRI, F. & CAMERON, A. G. W. 1974 *Icarus* **22**, 416.

PICKETT, B. K., MEJÍA, A. C., DURISEN, R. H., CASSEN, P. M., BERRY, D. K., & LINK, R. P. 2003 *ApJ* **590**, 1060.

PODOLAK, M. 2003 *Icarus* **165**, 428.

PODOLAK, M., POLLACK, J. B., & REYNOLDS, R. T. 1988 *Icarus* **73**, 163.

POLLACK, J. B. 1985. In *Protostars and Planets II* (Eds. D. C. Black & M. S. Matthews). p. 791. Univ. Arizona Press.

POLLACK, J. B. & BODENHEIMER, P. 1989. In *Origin and Evolution of Planetary and Satellite Atmospheres* (eds. S. K. Atreya, J. B. Pollack, & M. S. Matthews). p. 564. Univ. Arizona Press.

POLLACK, J. B., HUBICKYJ, O., BODENHEIMER, P., LISSAUER, J. J., PODOLAK, M., & GREENZWEIG, Y. 1996 *Icarus* **124**, 62. (Paper 1).

POLLACK, J. B., MCKAY, P., & CHRISTOFFERSON, B. 1985 *Icarus* **64**, 471.

RICE, W. K. M. & ARMITAGE, P. J. 2003 *ApJ* **598**, L55.

RICE, W. K. M., ARMITAGE, P. J., BATE, M. R., & BONNELL, I. A. 2003 *MNRAS* **339**, 1025.

RICHARDSON, L. J., DEMING, D., & SEAGER, S. 2004. In *Extrasolar Planets: Today and Tomorrow* (eds. J.-P. Beaulieu, A. Lecavelier des Etangs, & C. Terquem). ASP Conf. Ser. 321, p. 211. ASP.

RUSSELL, H. N. 1935 *The Solar System and Its Origin*. Macmillan.

SAFRONOV, V. S. 1969 *Evolution of the Protoplanetary Cloud and Formation of the Earth and Planets*. Nauka Press; English translation: NASA-TTF-677, 1972.

SANTOS, N. C., ISRAELIAN, G., & MAYOR, M. 2000 *A&A* **363**, 228.

SANTOS, N. C., ISRAELIAN, G., & MAYOR, M. 2001 *A&A* **373**, 1019.

SANTOS, N. C., ISRAELIAN, G., & MAYOR, M. 2004 *A&A* **415**, 1153.

SANTOS, N. C., ISRAELIAN, G., MAYOR, M., REBOLO, R., & UDRY, S. 2003 *A&A* **398**, 363.

SAUMON, D., CHABRIER, G., & VAN HORN, H. M. 1995 *ApJS* **99**, 713.

SAUMON, D. & GUILLOT, T. 2004 *ApJ* **609**, 1170.

SHAKURA, N. I. & SUNYAEV, R. A. 1973 *A&A* **24**, 337.

SHMIDT, O. 1944 *C.R. Dokl. Acad. Sci. URSS* **45**, 229.

SIGURDSSON, S., RICHTER, H. B., HANSEN, B. M., STAIRS, I. H., & THORSETT, S. E. 2003 *Science* **301**, 193.

SOZZETTI, A. 2004 *MNRAS* **354**, 1194.

STROM, S. E., EDWARDS, S., &SKRUTSKIE, M. F. 1993. In *Protostars and Planets III* (Eds. E. H. Levy & J. I. Lunine). p. 837. Univ. Arizona Press.

TAJIMA, N. & NAKAGAWA, Y. 1997 *Icarus* **126**, 282.

TANAKA, H., TAKEUCHI, T., & WARD, W. R. 2002 *ApJ* **565**, 1257.

THOMMES, E. W., DUNCAN, M. J., & LEVISON, H. F. 2003 *Icarus* **161**, 431.

UREY, H. C. 1951 *Geochim. Cosmochim. Acta* **1**, 209; **2**, 263.

WARD, W. R. 1997 *ApJ* **482**, L211.

WEIZSÄCKER, C. F. VON 1944 *Z. Astrophys.* **22**, 319; translation, Report No. RSIC-138 [=AD-4432290]. Redstone Scientific Information Center.

WETHERILL, G. W. 1980. *ARAA* **18**, 77.

WUCHTERL, G. 1991a *Icarus* **91**, 39.

WUCHTERL, G. 1991b *Icarus* **91**, 53.

WUCHTERL, G. 1993 *Icarus* **106**, 323.

WUCHTERL, G. 1995 *Earth, Moon and Planets* **67**, 51.

WUCHTERL, G. 1999. In *Theory and Tests of Convection in Stellar Structure* (Eds. A. Gimènez, E. F. Guinan, & B. Montesinos). ASP Conf. Ser. 173, p. 185. ASP.

WUCHTERL, G., GUILLOT, T., & LISSAUER, J. J. 2000. In *Protostars and Planets IV* (Eds. V. Mannings, A. P. Boss, & S. Russell). p. 1081. Univ. Arizona Press.

YOUNG, R. E. 2003 *New Astronomy Reviews* **47**, 1.

# Gravitational instabilities in protoplanetary disks

## By RICHARD H. DURISEN

Department of Astronomy, Indiana University, 727 E. 3rd Street, Bloomington, IN 47405-7105,
USA

In a protoplanetary disk that is sufficiently cold and massive, gravitational instabilities (GIs) will lead to the development of dense spiral waves on a dynamic time scale. For sufficiently short cooling times, comparable to about half a rotation period, an unstable disk will fragment into dense clumps that could be the precursors of gas giant protoplanets. At moderate cooling rates, the strong spiral waves which permeate the disk do not fragment, but nevertheless generate significant mass and angular momentum transport. I will review recent research on GIs with an emphasis on several critical questions: Do GIs cause planets to form? How fast do they transport mass? When do they occur? How do they affect the solids in the disk? The physical processes that are central to answering these questions are radiative and possibly convective cooling, irradiation of the disk, and gas-solid interactions. I conclude that, while it is unlikely that gas giant planets are formed directly by disk instability, GIs may substantially accelerate both planetesimal formation and core accretion.

## 1. Introduction

"These conjectures on the formation of the stars and the solar system I present with all the distrust which everything which is not a result of observations or of calculations ought to inspire." —Pierre Simon de Laplace, 1796.

### 1.1. *Focus*

Although we now have sufficiently conclusive observational evidence that the Kant-Laplace Nebular Hypothesis for the origin of planetary systems is correct in broad outline, and although we have much greater theoretical capacity to analyze the relevant physical processes than was available to Kant and Laplace, we are still far from achieving a consensus about how planets form from protoplanetary disks. Since Boss (1997) revived the Kuiper (1951) and Cameron (1978) idea that gas giant planets might form all at once through disk instability triggered by self-gravity, the planetary science and astrophysics communities have been embroiled in a stimulating debate about the relative merits of the disk instability mechanism and the "standard" core accretion plus gas capture picture of gas giant planet formation (Hubickyj, this volume). Much of the debate has hinged on how fast gas giant planet formation must happen due to finite gas disk lifetimes versus how fast core accretion can proceed. In this review, I will take a different tack. Whether or not gravitational instabilities (GIs) are involved in planet building, they are likely to occur in disks formed around stars by the collapse of rotating interstellar clouds (e.g., Laughlin & Bodenheimer 1994; Yorke & Bodenheimer 1999). This chapter concentrates on GIs as a physical process and attempts to bring the reader up to date on what is known about how disks behave when they become unstable (for earlier reviews, see Durisen 2001; Durisen et al. 2003). Although GIs in particulate disks and subdisks are another topic of contemporary interest (Youdin & Shu 2002), I will discuss only gas dynamical GIs.

### 1.2. *The big questions and an outline*

There are several important questions and attendant sub-questions that will be addressed in this review:

- Do gravitational instabilities produce gas giant planets?
  - Do they form planets directly on a dynamic time scale?
  - Or do they accelerate core accretion, instead?
- How fast can GIs transport mass in a disk?
  - Is this process local or global?
  - Does it produce persistent structures, like dense arms and rings?
- When do GIs occur in disks?
  - Do they occur in the early embedded phase?
  - Do they occur in dead zones, where turbulent transport by other mechanisms breaks down?
- How do GIs affect solids?
  - Do they mix solids?
  - Do they concentrate solids into coherent structures?
  - Do they cause thermal processing of solids?

I begin in Section 2 by describing the general characteristics of GIs that most, if not all, researchers currently agree upon. What emerges is unanimity about the central importance of radiative cooling for understanding the strength and outcome of GIs. As a result, the next two sections present in detail recent results from work with simple idealized cooling laws (§3) and realistic radiative cooling (§4). Additional physical processes, such as hydraulic jumps, gas-solid interactions, and the effect on GIs of other transport mechanisms, are discussed in Section 5.

Section 6 returns to the big questions posed above and presents some tentative answers.

Some of us, like myself, are visual thinkers, so I have sprinkled this review liberally with images. Although snapshots are helpful, movies and animations are sometimes even more informative when trying to grasp 3D flows. A number of relevant animations based on simulations by my own hydrodynamics group are available under the "Movie" tab at http://westworld.astro.indiana.edu/.

## 2. Secure foundations

### 2.1. *Linear regime*

Toomre (1964) showed analytically that gravitational instabilities to ring-like modes occur in thin gas disks when the parameter $Q = c_s\kappa/\pi G\Sigma$ is less than unity. Here $c_s$ is the sound speed, $\kappa$ is the epicyclic frequency at which a fluid element perturbed from circular motion will oscillate, $G$ is the gravitational constant, and $\Sigma$ is the surface density. High pressure, represented by $c_s$, stabilizes short wavelengths, and a high rotation rate, represented by $\kappa$, stabilizes long wavelengths. For a Keplerian disk, $\kappa = $ the rotational angular speed $\Omega$. The $\Sigma$ in the denominator conveys the destabilizing effect of disk self-gravity. A large body of numerical simulations, including those referenced in this review and dating back at least to Papaloizou & Savonije (1991), show that nonaxisymmetric modes leading to spiral structure are unstable for $Q \lesssim 1.5$–1.7. The instability is linear and dynamic, which means that small perturbations grow exponentially on the time scale of a rotation period $P_{\rm rot} = 2\pi/\Omega$ (e.g., Laughlin et al. 1998; Pickett et al. 1998). The precise $Q$-limit for instability depends somewhat on the structure of the disk. The growing multi-arm spirals have a predominantly trailing pattern, the amplification mechanism appears to be swing, and multiple modes with different numbers of arms can grow simultaneously (Nelson et al. 1998; Pickett et al. 1998; Laughlin et al. 1998; Mayer et al. 2004). Because the critical $Q$ for nonaxisymmetric modes is higher than that for axisymmetric modes, these should be the ones encountered in Nature.

## 2.2. *Nonlinear regime*

The behavior of GIs in the nonlinear regime is difficult to treat analytically, and so the results summarized below come almost entirely from simulations. Recent numerical calculations have differed in geometry and hydrodynamic algorithm. Examples include high-order governing equations in 3D (Laughlin et al. 1998), 2D (thin disk) shearing box local simulations (Gammie 2001), 2D (Nelson et al. 1998), and 3D global grid-based simulations in spherical (Boss 1997 to 2005) and cylindrical (Pickett et al. 1998, 2000, 2003; Mejía et al. 2005a) coordinates, and 2D (Nelson et al. 1998, 2000) and 3D (Mayer et al. 2002, 2004; Rice et al. 2003; Lodato & Rice 2004, 2005) Smoothed Particle Hydrodynamics (SPH) simulations.

As GIs reach nonlinear amplitudes, two major effects control their further development. The first and most important is the balance achieved between the loss of energy by radiative cooling and heating of the disk by dissipation of energy associated with the spiral waves. In many calculations, the latter takes the form of shock heating. That a balance of heating and cooling would control the nonlinear behavior of GIs was anticipated by Goldreich & Lynden-Bell (1965), and this notion has been used as a basis for developing accretion disk models for GI-active disks (e.g., Paczyński 1978; Lin & Pringle 1987). Beginning with Tomley et al. (1991), thermal regulation of GIs has been confirmed by many researchers (Pickett et al. 1998, 2000; Nelson et al. 2000; Gammie 2001; Boss 2001, 2002b, 2003; Rice et al. 2003; Lodato & Rice 2004, 2005; Mejía et al. 2005a). The key parameter controlling the nonlinear amplitude is the cooling rate, i.e., how fast the disk is able to lose the thermal energy pumped into it by GIs. The source of the heating is ultimately gravitational energy. Some comes from the collapse or contraction of material into dense self-gravitating structures, but most is gravitational energy released by net mass transport within the disk.

The second effect that is important at large amplitudes is nonlinear mode coupling, studied extensively by Laughlin and his collaborators (Laughlin et al. 1997, 1998). Power quickly becomes distributed over modes with various wavelengths and number of arms, resulting in a self-gravitating turbulence or *gravitoturbulence*, in which gravitational torques and Reynold's stresses can be important on a range of scales (Gammie 2001; Lodato & Rice 2004).

Prior to the 1990s, most work on GIs was done in a thin-disk approximation, where the disk is assumed to be hydrostatic in the $z$-direction. A significant development over the past decade is the recognition that the vertical structure of the disk plays a crucial role, both for cooling and for essential aspects of the dynamics. This is strongly emphasized by Pickett et al. (1998, 2000, 2003), who note an apparent relationship between the spiral modes in disks and the surface or $f$-modes of stars (see also Pickett et al. 1996; Lubow & Ogilvie 1998). In 3D, the notion of spiral "density" waves is only truly applicable to vertically isothermal disks, where the mode amplitude is actually uniform with height. Otherwise, GIs characteristically have large amplitudes at the surface of the disk. Although shock compression occurs in GI waves, it is accompanied, in disks with vertical temperature stratification, by extremely large surface distortions, strong vertical motions, and disproportionately greater shock heating at high disk altitudes. A dramatic illustration of such behavior will be presented in Section 5.1.

## 2.3. *Fragmentation criteria*

A consensus answer has emerged over the last few years about when turbulent GI spiral structure may fragment into discrete dense pieces. Let the cooling time $t_{cool}$ be defined as the gas internal energy density $\epsilon$ divided by the volumetric cooling rate $\Lambda$. For power-law equations of state and with $t_{cool}$ prescribed to be some value over an annulus of the disk,

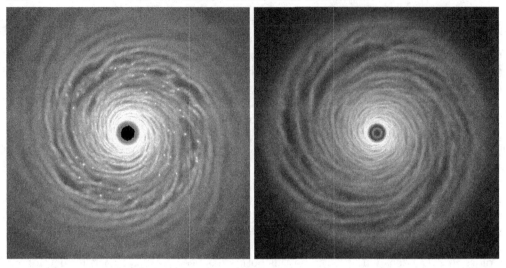

FIGURE 1. Equatorial plane density structures for SPH simulations of the same disk with different cooling times. *Left.* With $t_{\rm cool}\Omega = 3$, the disk fragments into dense clumps. *Right.* With $t_{\rm cool}\Omega = 5$, no fragmentation occurs. Figures are adapted from Rice et al. (2003).

the thin shearing box simulations of Gammie (2001) show that fragmentation occurs if and only if $t_{\rm cool}\Omega \lesssim 3$, or, equivalently $t_{\rm cool} \lesssim P_{\rm rot}/2$, where $P_{\rm rot}$ is the disk rotation period. This is confirmed in global SPH simulations by Rice et al. (2003), as shown in Figure 1. The disk in Figure 1 has a disk-to-star mass ratio $M_d/M_s = 0.1$, surface density $\Sigma \sim r^{-7/4}$, and an initial $Q = 2$. When evolved with $t_{\rm cool}\Omega = 3$, it fragments, but, when $t_{\rm cool}\Omega = 5$, it does not. Grid-based calculations by Mejía et al. (2005a) also yield fragmentation results consistent with Gammie (2001). The critical value of $t_{\rm cool}\Omega$ can be somewhat larger than three for more massive and physically thicker disks (Rice et al. 2003). For disks evolved under isothermal conditions, in the sense either that the fluid elements are forced to maintain the same temperature or that the disk temperature is kept constant locally at its initial value, a simple cooling time cannot be defined. In this case, thin shearing box simulations by Johnson & Gammie (2003) give fragmentation if and only if $Q < 1.4$. This agrees roughly with results from global simulations by other groups taken as a whole (e.g., Boss 2000; Nelson et al. 1998; Pickett et al. 2000, 2003; Mayer et al. 2002, 2004). Classic examples of isothermal disk fragmentation are shown in Figure 2.

So, all researchers agree that a disk will fragment into dense clumps if it is sufficiently cool and evolves isothermally, or if it cools on a short time scale compared with a rotation period. Where controversy remains is how long-lived these clumps may be. Because the peak clump densities always increase with increasing numerical resolution, and because they always seem to have thermal energies dominated by self-gravitational energies, Boss (2005) concludes that they are permanent bound objects. On the other hand, Pickett et al. (2003) and Mejía et al. (2005a) find, regardless of resolution, that clumps are eventually destroyed by tidal stresses, shears, and collisions. Unfortunately, except for the case shown in Figure 2, Boss's highest-resolution simulations (especially Boss 2005) are usually not integrated for long times, so it remains unclear whether or not his clumps might eventually be disrupted.

Figure 3 presents a sobering result on clump longevity obtained with the Pickett et al. (2003) code. The same low-$Q$ disk is evolved isothermally at high resolution with and without artificial bulk viscosity (ABV) in the momentum equation. Without ABV (left

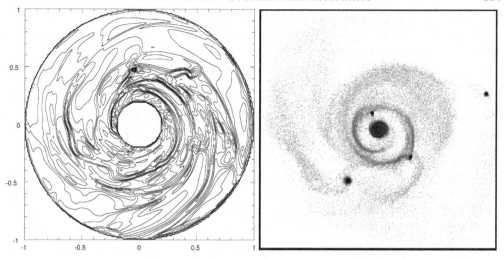

FIGURE 2. Equatorial densities for two simulations of isothermally evolved disks that fragment into long-lived multi-Jupiter mass clumps. *Left.* This grid-based calculation by Boss (2000) shows a 0.9 $M_\odot$ disk with a 20 AU outer radius and a minimum $Q \sim 1.3$ orbiting a 1 $M_\odot$ star after several outer rotations. The dense clump near 12 o'clock can be tracked for many orbits. *Right.* Long-lived clumps form in an SPH simulation by Mayer et al. (2004) for a 0.1 $M_\odot$ and $Q = 1.38$ disk with an initial outer radius of 20 AU orbiting a 1 $M_\odot$ star. Figures are adapted from Boss (2001) and from Mayer et al. (2004).

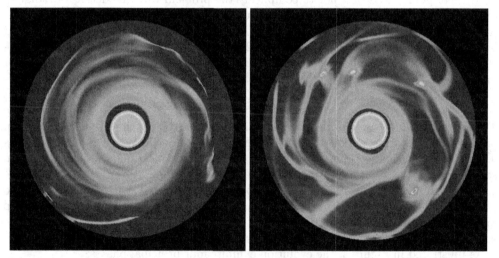

FIGURE 3. Midplane density grayscales for a Pickett et al. (2000) star/disk model with $Q = 1.35$ evolved at high resolution (512 azimuthal zones) with the mass distribution of the central star frozen. *Left.* Without artificial bulk viscosity, fragments form within one outer rotation, but have all been destroyed through violent interactions by 4.9 outer rotations, the time shown. *Right.* With artificial bulk viscosity, dense fragments still survive at the same evolutionary time as shown in the left panel. The fragment closest to 45 degrees left of the vertical can be traced for at least four rotations, and persists until the simulation is stopped at 7.3 outer rotations. Figures are provided courtesy of M. K. Pickett.

panel), the disk fragments rapidly into many clumps, but the violent interactions of the clumps destroys them all by the time shown in Figure 3. With ABV on in the momentum equation (right panel), clump formation is delayed and is less violent, so that the clumps, once formed, are long lived. One clump in the right panel of Figure 3 can be traced for four complete orbits, and still exists when the simulation is stopped. The point of this demonstration is that clump longevity is sensitive to details of the numerical treatment. ABV is an unavoidable feature of SPH simulations where long-lived clumps are most commonly seen (Mayer et al. 2004). However, most of the simulations by Boss do not have an explicit ABV, and so it is unclear whether Figure 3 has any relevance to clump survival in his code. To complicate matters further, fragmentation can occur for purely numerical reasons (Bate & Burkert 1997; Truelove et al. 1997; Nelson 2003).

It will undoubtedly require a great deal of future effort to determine whether fragments in disks are truly long-lived protoplanets. It may prove more profitable to ascertain instead whether "real" disks ever cool fast enough to fragment in the first place, and to investigate other important effects related to GIs and planet formation that are more tractable. The rest of this review focuses on several recent advances in our understanding of GIs, particularly regarding how GIs interact with, and are affected by, a range of other physical processes, including radiative cooling.

## 3. Idealized cooling

In this section, I discuss how disks with GIs evolve when they do not fragment. An important question in this context is whether the evolution can be adequately approximated by a local $\alpha$-disk prescription (e.g., Lin & Pringle 1987) or whether GIs are intrinsically a global phenomenon (Laughlin & Rozyczka 1996; Balbus & Papaloizou 1999). To address these points, it is best not to complicate the problem with the details of radiative transport, but to examine the results of simulations with simple, idealized cooling laws. A body of simulations exists where cooling by the disk is assumed to be characterized by requiring that either $t_{cool}$ (Pickett et al. 2003; Mejía et al. 2005a) or $t_{cool}\Omega$ (Gammie 2001; Rice et al. 2003; Lodato & Rice 2004, 2005) has the same constant value everywhere in the disk.

### 3.1. $t_{cool} = constant$

The longest simulation performed with a constant global cooling time is the $t_{cool} = 2$ orp (orp = outer rotation period) 3D simulation described by Mejía et al. (2005a) for a 0.07 $M_\odot$ disk stretching from 3 to 40 AU (AU = astronomical unit) around a 0.5 $M_\odot$ star. The initial equilibrium disk has a surface density profile $\Sigma \sim r^{-1/2}$ except at the edges, is marginally stable to GIs with a minimum $Q$ of about 1.8 at $r = 30$ AU prior to the beginning of cooling, and has a meridional aspect ratio of about ten to one. The orp for this disk is defined as the initial rotation period at 33 AU or about 250 yrs. The disk is peppered initially with low amplitude random perturbations to its density.

As illustrated in Figure 4, the evolution exhibits four principal phases:

1. During the *cooling phase* (not shown), which lasts only a few orps, the disk cools to the instability point and becomes vertically thinner. The disk remains axisymmetric in appearance, but perturbations begin to grow in the outer disk where $Q$ is lowest.

2. Between about four to eight orps, during the *burst phase* (left panel), a single well-defined multi-armed spiral mode develops, becomes extremely nonlinear, and produces a large pulse of mass redistribution. Spiral arms are ejected outward in a manner similar to the dynamic bar-like instability of rotating stars (Durisen et al. 1986).

3. When these arms fall back partially, very strong shocks develop which heat the disk. It then enters a *transition phase*, lasting from about 8 to 12 orps. As the disk becomes hotter, $Q$ rises and the nonaxisymmetry washes out.

4. Cooling reasserts itself by 12 orps, and the disk settles into a long-lived *asymptotic phase* with an overall balance between heating and cooling. Persistent GIs manifest

 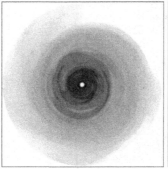 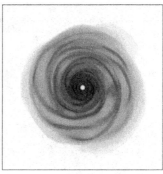

FIGURE 4. Equatorial mass density grayscales for the Mejía et al. (2005a) disk evolved with $t_{cool} = 2$. *Left.* The Burst Phase at $t = 5$ orps (outer rotation periods). Strong instability occurs from four to eight orps in a discrete four-armed mode which rapidly redistributes mass in the disk. *Center.* The Transition Phase at ten orps. Shock heating due to the burst heats the disk, and nonaxisymmetric structure temporarily washes out between 8 and 12 orps. *Right.* The Asymptotic Phase at $t = 20$ orps. By about 12 orps, the disk achieves a long-term balance between heating and cooling with sustained self-gravitating turbulence. The panels in this figure are about 170 AU along an edge. Figures are adapted from Mejía et al. (2005a).

themselves as a complex nonlinear system of numerous multi-armed spirals. During the asymptotic phase, the disk maintains a nearly constant value of $Q \approx 1.45$ over $r = \sim12$–40 AU, but with spatial and temporal fluctuations about this mean. This phase continues without significant qualitative change until the calculation is stopped arbitrarily at 23.5 orps (5,875 yrs).

Radially inward mass transport peaks at $\sim10^{-5}\ M_{\odot}/yr$ in the burst phase, a bit shy of FU Orionis outburst values (Bell et al. 2000). Throughout the asymptotic phase, the disk sustains a highly variable inflow averaging about $5 \times 10^{-7}\ M_{\odot}/yr$ over the radial range of 10 to 30 AU with a similar outflow rate beyond 30 AU. Fourier decomposition of the density structure into $\sin(m\phi)$ and $\cos(m\phi)$ components, where $m$ corresponds to the number of arms in a spiral disturbance, provides some insight into the source of the transport. For all $m$'s tested ($m = 1$ to $6$) except $m = 1$, there are dozens of coherent modes throughout the disk that span the radii between their inner and outer Lindblad resonances. Although not obvious to the eye in the right panel of Figure 4, there appear to be persistent two-armed ($m = 2$) modes with corotation radii (CR) near 30 AU during the 11.5 orps of the asymptotic phase that we followed. As expected for gravitational torques exerted by a trailing spiral mode, we see inward transport inside the CR of these two-armed modes and outward transport outside the CR. We conclude that the mass and angular momentum transport is dominated in this case by a few low-order global modes, and direct computation of the gravitational stresses confirms it.

A remarkable feature of this simulation is the growth of a series of rings near the boundary between the outer GI-active disk ($r > 10$ AU) and an inner disk that remains too hot to sustain GIs on it own. Figure 5 demonstrates the ring growth. The cause of this phenomenon and its potential significance for planet formation will be discussed in Section 5.1.

It is natural to wonder how these results depend on the somewhat arbitrary choices of simulation parameters. In Mejía et al. (2005a), when the 12 to 18 orp stretch is repeated with $t_{cool} = 1$ orp, the disk again achieves an overall balance of heating and cooling with a nearly constant $Q \approx 1.45$, but the GIs become stronger, as evidenced by an increase in the Fourier amplitudes of the spirals, a doubling of the mass inflow rate, and faster growth of the rings in the inner disk. A comparison of mass inflow rates for $t_{cool} = 0.25$,

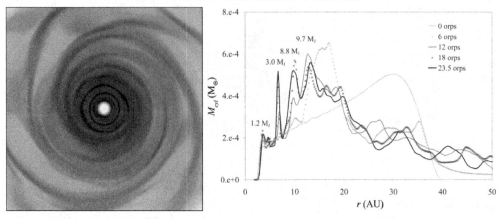

FIGURE 5. The development of dense rings in the inner disk of a $t_{\rm cool} = 2$ orp simulation. *Left.* This equatorial plane density grayscale shows the rings in the innermost disk region near the end of the simulation at 23.5 orps. The panel is about 85 AU along an edge. *Right.* The mass contained within a cylindrical shell one-sixth of an AU in width is plotted as a function of radius to illustrate the growth of multiple Jupiter mass concentrations of mass. The features at 7 and 10 AU are truly ring-like, while the one at 13 AU has active nonaxisymmetric structure. Figures are provided courtesy of A. C. Mejía. The left panel is from Mejía et al. (2005a).

1, and 2 orp runs shows that mass inflow rates during the asymptotic phase scale as $t_{\rm cool}^{-1}$, in agreement with Tomley et al. (1991). My hydrodynamics group has recently completed two additional $t_{\rm cool} = 2$ orp runs where the initial disks have $\Sigma \sim r^{-a}$, where $a = 1$ and $3/2$. These evolutions exhibit the same four phases and other behaviors as for $a = 1/2$, except that the burst weakens and the rings become more prominent (Michael & Boley, private communication).

### 3.2. $t_{\rm cool}\Omega = constant$

Lodato & Rice (2004, 2005) analyze how GIs behave for a range of $M_d/M_s$ from 0.05 to 1.0 when $t_{\rm cool}\Omega$ is set to the nonfragmenting value of 7.5. Their disk has an initial $\Sigma \sim r^{-1}$ and a temperature structure $T \sim r^{-1/2}$, giving a minimum initial $Q \approx 2$. The choice $t_{\rm cool}\Omega = 7.5$, equivalently $t_{\rm cool}/P_{\rm rot} \approx 1.2$, pins the cooling time everywhere to the local rotation period. For low $M_d/M_s$, Lodato & Rice (2004) find that the evolution of the disk tends to be self-similar. After the disk achieves balance between heating and cooling, tightly wrapped spirals exist everywhere in the disk with similar pitch angles, and the disk settles into sustained instability with $Q \sim 1$ to 1.5, as illustrated in the left panel of Figure 6. No pronounced "burst" is reported, and the stresses and resultant transport produced by the GI activity is reasonably well described as a local phenomenon.

However, for the vertically thicker and more massive disks, especially with $M_d/M_s = 0.5$ and 1.0, Lodato & Rice (2005) obtain evolutionary phases similar to those described by Mejía et al. (2005a), with a strong burst in a global mode that transitions into an asymptotic phase. As $M_d/M_s$ increases, the amplitudes of low-order (few-armed) modes in the asymptotic gravitoturbulence become more important relative to the higher-order modes. With a wide radial range spanned by the region between their Lindblad resonances, low-order modes communicate torques in a global way. So, very massive disks ($M_d/M_s \geqslant 0.25$) exhibit some global behavior even when their cooling times are forced to be local. For $M_d/M_s = 1.0$, the disk continues to experience burst-like episodes of global mode behavior rather than settling into an asymptotic phase.

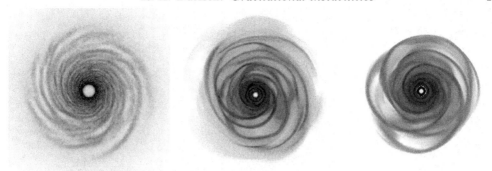

FIGURE 6. Equatorial plane density grayscales at late times. *Left.* The $M_d/M_s = 0.1$ disk with $t_{cool} = 1.2P_{rot}$ and initial $\Sigma(r) \sim r^{-1}$ from Lodato & Rice (2004) has a smooth self-similar appearance with tight spirals. *Center.* The $t_{cool} = 1$ orp simulation with $M_d/M_s = 0.14$ and an initial $\Sigma \sim r^{-1/2}$ from Mejía et al. (2005a) has more open, global spirals. *Right.* This disk is similar to the one in the center panel except that it has an initial $\Sigma \sim r^{-1}$ and $t_{cool} = 2$ orps. All figures have a similar radial scale. Figures are provided courtesy of W. K. Rice, A. C. Mejía, S. Michael, and A. C. Boley.

Lodato & Rice compute the gravitational and hydrodynamic (Reynolds) stresses induced by GIs in their disks. Focusing on the inner to middle disk regions, where there are negative net torques and radial inflow of mass, Lodato & Rice compare the stresses with a prediction by Gammie (2001) that the effective $\alpha$ viscosity should be well approximated by

$$\alpha = \left| \frac{d\ln\Omega}{d\ln r} \right|^{-2} \frac{1}{\gamma(\gamma - 1)t_{cool}\Omega} \, . \tag{3.1}$$

If equation (3.1) were correct, it would allow long-term evolutions of the radial mass distribution in GI-active disks to be done with a simple $\alpha$-viscosity prescription (e.g., Yorke & Bodenheimer 1999; Armitage et al. 2001). Contrary to Gammie's thin shearing box local simulations, where he finds that the gravitational and Reynolds stresses are similar, Lodato & Rice always find that gravitational stresses dominate. Nevertheless, for $t_{cool}\Omega = 7.5$, equation (3.1) predicts $\alpha = 0.05$, which is close to the typical effective $\alpha$ that Lodato & Rice detect in the inflow regions of their disks. This suggests that, on average, equation (3.1) could be a useful approximation for $\alpha$-disk modeling of GIs. There are, however, substantial ($\sim$10%) radial and, presumably, temporal variations about this value. Equation (3.1) is not useful for characterizing the global burst when $M_d/M_s$ is large, and deviations from (3.1) do seem systematically larger as $M_d/M_s$ increases. However, Lodato & Rice find no evidence at any $M_d/M_s$ for significant wave-transport of energy expected when GIs behave globally (Balbus & Papaloizou 1999). Local heating rates seem to match local cooling rates throughout their disks.

### 3.3. *Conclusions*

Apparently, when cooling is treated as set by a global time scale ($t_{cool} = $ constant), as in Mejía et al. (2005a), GIs behave in a global manner, whereas, when cooling is treated as a locally determined phenomenon ($t_{cool}\Omega = $ constant), as in Lodato & Rice (2004, 2005), GIs behave locally, unless $M_d/M_s$ is large. For $t_{cool}\Omega = $ constant, a local effective $\alpha$ prescription apparently fits the GI stresses in simulations fairly well. This result is perhaps not too surprising if we recall that $t_{cool}\Omega = $ constant is a necessary condition for a steady-state accretion disk. With $t_{cool} = $ constant, on the other hand, Mejía et al. (2005a) find that the effective $\alpha$ for the average inflow during the asymptotic phase

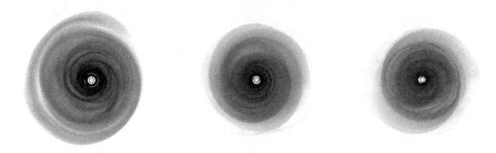

FIGURE 7. Equatorial plane density grayscales for three calculations with radiative cooling during the asymptotic phase with varying amounts of envelope irradiation. *Left.* No irradiation. *Center.* $T_{\text{irr}} = 15$ K. *Right.* $T_{\text{irr}} = 25$ K. The initial equilibrium model is the same as for the simulation in Figure 4. The width of each panel is about 140 AU. Figures are provided courtesy of K. Cai.

is a few to many tens of times larger than given by equation (3.1) and is extremely variable, even changing sign for intervals of time. Figure 7 shows a direct comparison of disks evolved under the two types of cooling law. Even visually, it is apparent that the $t_{\text{cool}}\Omega$ = constant disk on the left is more self-similar, while the spirals in the right two panels tend to be more open and irregular with more radially variable structure, including dense rings in the innermost disk.

To summarize all this in one sentence, simulations to date show that, for moderate disk masses, whether GIs are a local or global phenomenon depends on whether the cooling time itself is local or global. As in Pickett et al. (1998, 2000), the lesson is that the behavior of GIs is controlled by the details of the thermal physics. So, let us now consider whether realistic treatments of radiative cooling lead to cooling times that can be characterized as local or global.

## 4. Radiative cooling

### 4.1. *Approaches*

In pioneering work by Nelson et al. (2000), the midplane temperatures and surface densities from 2D thin-disk hydrodynamics simulations are used to fit simple vertical structures in hydrostatic equilibrium. The vertical fits then permit an estimate of the cooling rate for each column based on realistic opacities. In several simulations of Solar Nebula-sized disks, disk fragmentation, if it occurs at all, is much reduced over isothermal evolutions of the same disks. As an extension of Gammie's thin shearing box simulations, Johnson & Gammie (2003) use a similar approach by fitting what is effectively a "one-zone" Eddington-like vertical radiative equilibrium solution to every midplane cell of their 2D disks. Because their treatment is local, they are able to consider a wider range of parameter space than Nelson et al. (2000). They find that the Gammie fragmentation criterion applied to the initial cooling time is not a good measure for the likelihood of fragmentation, especially in parameter regimes where the opacity varies significantly with temperature due to evaporation of grain types. An average cooling time evaluated after the GIs become nonlinear is more reliable.

So far, there are only two treatments of radiative cooling for protoplanetary disks in full 3D for which I have sufficient information to review—those by Alan P. Boss (2001, 2002a,b, 2004a, 2005) and those by members of my own hydrodynamics group, especially

Annie C. Mejía (2004, 2005b) and Kai Cai (Cai et al. 2006). Although the work by Mejía and Cai is not yet published, radiative cooling is so important for understanding the behavior of GIs in real disks that I will unveil some interesting results that are currently in preparation. Several other research teams are working on this problem (e.g., Mayer, private communication), but I am not privy to any detailed preliminary information. Another student in my group, Aaron C. Boley, is collaborating with Åke Norlund to adapt a ray scheme for 3D radiative transfer (Heinemann et al. 2006) to disk geometry, but no disk simulation results are yet available. Because there are serious disagreements about the effects of radiative cooling between Boss and Mejía/Cai, it is worthwhile to lay out methodological similarities and differences in some detail.

### 4.2. *Boss versus Mejía/Cai methodologies and results*

Both Boss and Mejía/Cai use the same algorithm to do flux-limited diffusion in 3D within the optically thick part of the disk (Bodenheimer et al. 1990). The photosphere of the disk is determined by computing an optical depth $\tau$ either inward along the spherically radial direction (Boss) or downward along the cylindrically vertical direction (Mejía/Cai). Boss and Mejía/Cai differ primarily in how they treat the optically thin regions outside the photosphere and how they match the optically thick and thin regions at the photosphere. Boss basically resets his optically thin cells either to their initial temperature or to a background temperature. The background temperature is intended to represent a thermal bath due to IR irradiation from an infalling envelope (Chiang & Goldreich 1997; Chick & Cassen 1997). He does not attempt to match fluxes at the thin/thick boundary. Mejía/Cai solve the energy equation in optically thin regions, allowing the gas to radiate freely. Without coupling to the upward flowing radiation from the optically thick disk, however, regions above the photosphere cool precipitously and do not approach radiative equilibrium. To prevent this from happening, some heating by the upward-moving radiation is included. As a boundary condition for the optically thick diffusion, the thin and thick solutions are fit together, albeit crudely, over one or two cells near the photosphere by an Eddington gray atmosphere. The Mejía/Cai algorithm takes so much care with the optically thin regions for two main reasons. First, the fractional volume of the optically thin part of the disk often exceeds that of the optically thick part, and it tends to grow as the disk cools and opacities decrease. Second, disproportionately large heating rates at higher altitudes in the disk due to shocks and irradiation can damp GIs (§4.3), and so deposition and removal of energy from these regions must be modeled.

Let us summarize significant differences between calculations by Boss and by Mejía/Cai:

1. *Radiative Boundary Conditions (BCs) and the Optically Thin Regions*, as detailed in the preceeding paragraph.

2. *Equation of State.* Mejía/Cai use a $\gamma = 5/3$ monatomic gas throughout their disk, while Boss includes effects of the rotational levels and dissociation of molecular hydrogen.

3. *Opacities.* Boss uses Pollack et al. (1994) opacities with fixed dust grain sizes; Mejía/Cai adopt the same opacities used by Calvet et al. (1991) and D'Alessio et al. (1998), which allow variation of the maximum grain size.

4. *Artificial Bulk Viscosity.* Mejía/Cai always include ABV in their runs to simulate entropy production by shock waves, whereas Boss does not include ABV in the vast majority of his runs.

5. *3D Grid.* The Boss code has a spherical grid; Mejía/Cai use a modified version of the Pickett et al. (2003) code with a cylindrical grid.

6. *Initial Model.* There are differences between the initial models used by the groups in terms of temperature $T(r)$ and surface density distributions $\Sigma(r)$.

7. *Initial Perturbations*. Mejía/Cai use a low-amplitude random density perturbation to the axisymmetric initial model; Boss gives his disk a strong hit with two, three, and four-armed modes plus a smaller random component.

Unfortunately, when we compare results from the two groups for disks with similar $M_d$, $M_s$, $Q$, and radii, it is hard to imagine how the outcomes could be more disparate. In a series of papers over the last few years, Boss (2001, 2002a,b, 2004a, 2005) has presented simulations in which marginally unstable, radiative disks of Solar System size cool rapidly enough to fragment due to efficient convection. He finds this to be independent of metallicity and of IR irradiation by a circumstellar envelope. As Boss (2005) increases his azimuthal resolution, he tends to see the maximum density of clumps increase, which leads him to conclude that they are bound objects.

Mejía/Cai, on the other hand, generally find that, after a short adjustment period, cooling times become long, disks do not fragment, GIs' amplitudes increase with decreasing metallicity, and GIs tend to be damped by irradiation. Convection, if it occurs at all, seems to be localized and does not lead to rapid cooling. Cai, Boss, and myself have begun a collaborative effort to understand these severe differences by computing identical disks with the different methods using the same code. So far, we tentatively rule out items #2 to 7 above as the culprits. It seems probable, but not yet certain, that the primary cause of the differences is item #1, the treatment of the radiative BCs and the optically thin region. Of course, I am inclined to believe that my group's approach is the better one, but this is far from a firm conclusion at present. I will summarize some of the Mejía/Cai results (Mejía et al. 2005b; Cai et al. 2006) in the next section under the presumption that our simulations are more representative of real disk behavior, but the reader should remain respectfully skeptical until more groups present radiative hydrodynamics disk modeling in 3D.

### 4.3. *Irradiation, metallicity, and grain size*

Our first attempt to treat radiative cooling (Mejía 2004; Mejía et al. 2005b) is a simulation where the dust grains have a maximum radius of $a_{max} = 1$ micron, where there is no external source of radiation shining down on the disk, and where the initial disk model is chosen to be the same as for the $t_{cool} = 2$ orp simulation discussed in Section 3.1. We call this simulation "Shade," because the disk surface is not irradiated. Shade exhibits the same four phases of evolution—cooling, burst, transition, and asymptotic—as our global $t_{cool}$ = constant simulations, and mass transport also appears to be dominated by low-order GI modes. These characteristics are shared by all Mejía/Cai simulations, and the GIs appear to behave in a global manner.

An examination of $t_{cool}(r, t)$ throws considerable light on these results. Here, we define $t_{cool}$ in a column-wise sense, namely, $t_{cool}$ = the internal energy per unit projected disk area divided by the radiative flux out the top of the column. Initially, $t_{cool}$ is relatively low, an orp or less, and decreases with $r$. This $t_{cool}(r)$ is an artifact of our initial vertically isentropic disk structure, which is very far from radiative equilibrium. The temperature structure adjusts rapidly as the disk cools in such a way that $t_{cool}$ increases everywhere and becomes large, up to a few to 10 orps. It is relatively constant with $r$, although with very large local variations. Thus, after an initial radiative transient, Shade and all other Mejía/Cai simulations evolve toward a state best characterized as having a global $t_{cool}$. This situation is drastically different from the $t_{cool} \sim r^{3/2}$ required for $t_{cool}\Omega$ = constant. In this sense at least, GIs in real disks are a global, not local phenomenon. Typical mass inflow rates for Mejía/Cai disks in the asymptotic phase range between a few $\times 10^{-7}$ to somewhat more than $10^{-6}$ $M_\odot$/yr.

For reasons of computational cost, all Mejía/Cai calculations are run at an only moderate azimuthal resolution (128 cells from 0 to $2\pi$), probably insufficient to permit fragmentation if it wants to occur (Pickett et al. 2003). From experience, we think we know how a disk that would fragment at high resolution behaves at low resolution, but, to be sure, we do rerun some stretches of some calculations with much higher resolution (512 cells from 0 to $2\pi$). Although dense clumps do form during the burst of Shade, while the $t_{cool}$s are still short by the Gammie fragmentation criterion, the clumps do not persist for more than a fraction of an orbit. So far, although we have not tested all cases, we have seen no evidence for significant fragmentation in any Mejía/Cai simulation with radiative cooling. The $t_{cool}$s in the asymptotic phase, typically of the order of five orps, are above Gammie's critical value of $P_{rot}/2$ and are much higher than the initial $t_{cool}$s. We strongly agree with Johnson & Gammie (2003) that one cannot judge whether a disk will fragment by using *initial* values of $t_{cool}\Omega$.

### 4.3.1. *Irradiation*

Disks can be subject to irradiation by the central star (e.g., D'Alessio et al. 1999), by hot stars in a clustered environment (e.g., Johnstone et al. 1998), by a binary companion, and by a circumstellar envelope (e.g., Chiang & Goldreich 1997). The latter case is the easiest to handle, because an envelope will radiate in the far IR or at millimeter wavelengths where the opacities and optical depths are similar to those within our standard disk for $r >$ few AU. To test the effects of envelope irradiation, Cai et al. (2006) rerun the Shade calculation with an IR flux of blackbody temperature $T_{irr}$ shining down onto the disk from the grid boundary along the $z$-direction. Figure 7 compares the results at the same simulation time in the asymptotic phase for $T_{irr} = 0$ (Shade), 15 K, and 25 K. Irradiation suppresses the GI amplitudes. The difference between the 15 K and 25 K cases is a bit difficult to discern clearly in the figure. However, in extreme cases, for $T_{irr} >$ about 40 K, the disk is not able to cool, and $Q$ remains too large for instability to occur at all. Direct stellar irradiation is more difficult to treat, because the stellar photons are absorbed initially in a surface layer too thin for our code to resolve, but preliminary indications in Mejía (2004) are that it too tends to suppress GIs. Although other forms of irradiation have not yet been treated in GI simulations, I suggest that we will find that any external environmental factor that pumps heat into the disk will weaken if not prevent GIs. If this is correct, then comprehensive modeling of the radiative environment of real disks is required in order to determine whether GIs occur and how they behave.

### 4.3.2. *Metallicity*

Contrary to Boss (2002a, 2004a), Mejía/Cai find that metallicity matters. Cai et al. (2006) has repeated the $T_{irr} = 15$ K run with metallicities ranging from one quarter to twice the solar value. Equatorial plane density images from the two extreme cases are shown in the left and center panels of Figure 8. What the eye suggests when comparing Figure 7 (center panel) and Figure 8 (left and center panel) is confirmed by objective measures of GI amplitudes based on Fourier analyses, namely, GIs become stronger as the metallicity is lowered. Cai et al. (2006) do agree with Boss in one unexpected way. The optically thick parts of the disks converge to roughly similar behaviors and lose energy at similar rates regardless of metallicity. However, there is no efficient convection, and the cooling is slow, with $t_{cool}$ typically on the order of five to seven orps, except for localized regions which are now being tested for possible fragmentation at higher resolution. Major differences are found in the optically thin regions, which dominate the volume of the disk at low metallicity. If gas giant planets form directly, all at once, by

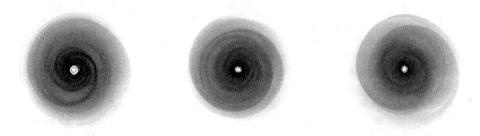

FIGURE 8. Equatorial plane density grayscales in the asymptotic state for three calculations with $T_{\mathrm{irr}} = 15$ K and varied metallicity and grain size. *Left.* One-quarter solar metallicity for $a_{\max} = 1$ micron. *Center.* Twice solar metallicity for $a_{\max} = 1$ micron. *Right.* Solar metallicity and $a_{\max} = 1$ mm. Images and simulations are similar in all other respects to those in Figure 7. Figures provided courtesy of K. Cai.

disk instability, then having stronger GIs at lower metallicity seems to run counter to the observed tendency for more planets to be found around high metallicity stars (Fischer & Valenti 2005). Anyhow, the $t_{\mathrm{cool}}$s generally seem too long for fragmentation to occur.

### 4.3.3. *Grain size*

Figure 8 shows that, at least in this disk model, use of a larger maximum grain size, $a_{\max} = 1$ mm, increases the opacity over the bulk of this rather cool disk, resulting in an effect similar to an increase in metallicity. Modeling of real disks does seem to require larger grain sizes (D'Alessio et al. 2001), even at early times (Osorio et al. 2003). The opacity variations are complex, however. In the hotter inner disk, increases in grain size and settling of grains should *reduce* rather than increase the opacity. Also, as shown by Nelson et al. (2000), as grains are transported by GIs through shocks and across temperature gradients, they may be vaporized. The coupling through opacity between GIs and grain growth, destruction, and settling will prove to be a serious challenge for future modeling. We will return to aspects of this in Section 5.

### 4.4. *Conclusions*

A great deal remains to be done before we claim to understand GIs in radiative disks with realistic opacities. Results available to date are almost as divergent as they possibly could be. Better treatments of both the optically thick and thin regions and their interface are required. If you believe the results of my own hydro group's calculations, then GIs are sensitive to their radiation environment and to variations in opacity. For the mass ($M_d/M_s = 0.14$) and size (few to 40 AU) of disk considered, cooling is, by and large, too slow under realistic conditions to produce fragmentation, and GIs can be weakened or even suppressed by external radiation fields impinging on disks. Similar conclusions have been reached recently by others through analytic arguments (Rafikov 2005; Metzner & Levin 2005). The reader must remember, however, that the most extensive body of published work on this subject, by Boss, argues for fast cooling by convection regardless of metallicity or envelope irradiation.

## 5. Special effects

Despite the sorry state of current affairs regarding radiative physics in unstable disks, significant progress is being made on other ways that GIs can affect the appearance and evolution of dusty gas disks.

### 5.1. *Hydraulic jumps*

As emphasized by Pickett et al. (1996, 1998, 2000, 2003), GIs produce large spiral corrugations and other more complex surface distortions in gas disks. Pursuant to a suggestion by D. Cox, A. C. Boley and I have been analyzing many of these vertical distortions as hybridized combinations of shocks plus hydraulic jumps, which we call hydraulic shock-jumps or *hs-jumps*. Martos & Cox (1998) showed that analogs of classic hydraulic jumps occur for compressible fluids in astrophysical disk geometry. Boley et al. (2005) and Boley & Durisen (2006) explain the jumps as follows. Consider the case where disk gas flows into the back of a trailing spiral wave with a pattern speed much slower than the gas orbital speed. Suppose the pre-shock gas is in vertical hydrostatic equilibrium. After passing through a strong adiabatic shock, the vertical pressure gradient of the post-shock gas can easily exceed any increase in vertical gravitational force, so that the gas accelerates upward. As it jumps above the pre-shock disk height, the gas also expands horizontally. This vertically and horizontally expanding plume also tends to stream inward along the spiral due to the reduction in the component of the gas velocity normal to the trailing spiral shock front. The net result is a wave curling back and inward over the spiral shock. Eventually, the vertically "jumping" gas crashes back down on the disk in a huge breaking wave along the spiral arm. The wave produces additional strong shocks at high altitudes in the disk and probably generates turbulence.

Picking this behavior out of the gravitoturbulence of a GI-active disk is daunting, and so Boley et al. (2005) and Boley & Durisen (2006) perform toy calculations that isolate the hs-jumps by stimulating simple well-defined spiral waves. An illustrative case is shown in Figure 9. Although the parameters are not particularly realistic, this simulation provides a clean case with a single two-armed wave. An initially axisymmetric disk is perturbed by a strong $\cos(2\phi)$ potential concentrated at about 5 AU and corotating with the fluid at 5 AU. The intention is to simulate what might happen in the asteroidal region if Jovian mass clumps suddenly appear, due perhaps to the eruption of GIs in a dead zone (see §5.3). The perturbation produces two corotating spiral shocks reaching well into the inner disk, as shown in the upper left panel of Figure 9. The upper right panel shows the corresponding surface disturbance. The curling and breaking character of the surface wave is best seen in the radial cut through the disk in the bottom panel. The vertical shaded region is the shock front along the inner edge of the spiral. The curling and breaking wave and its associated high altitude shocks are evident. Note that the scale of these structures is many tenths of an AU. Fluid elements traced in this flow suffer large radial and vertical excursions.

We are only beginning to explore the consequences of such wave action. For instance, Boley et al. (2005) find that shocks like those shown in Figure 9 have the characteristics, laid out by Desch & Connolly (2002), necessary to produce chondrules. Chondritic material seems to represent the bulk of the solids that condensed in the asteroid region, and the thermodynamics of its origin is a major cosmogonic puzzle. Following Wood (1996), Boss & Durisen (2005a,b) suggest that GI-generated shocks may have played *the* dominant role in the thermal processing and mixing of primitive material in the Solar Nebula. Time-varying surface distortions should also have observable effects in disks around young stars. In addition to photometric variability, unusual spectroscopic features could be produced as the surface corrugations lift otherwise shaded material into

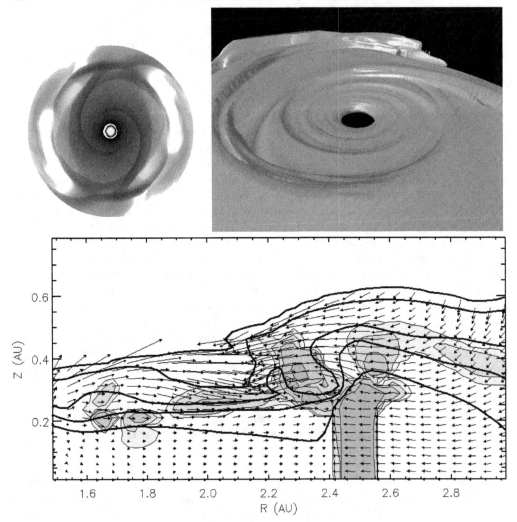

FIGURE 9. Strong spiral shocks produce hydraulic jumps in a massive inner Solar Nebula model. *Upper Left.* This equatorial plane density grayscale shows the two-armed shocks generated interior to a $\cos(2\phi)$ potential perturbation that rotates with the fluid at 5 AU and is concentrated at that radius. *Upper Right.* A 3D density contour viewed from above the disk. The density value for the contour is low enough to represent the disk surface. *Bottom.* This radial slice illustrates how the vertically jumping material curls inward over the main shock and crashes back down on the pre-shock gas in a breaking wave. The heavy solid lines are density contours, the light-shaded contours show shock heating, and the arrows are meridional velocity vectors of the gas. All figures are from the same time in the simulation. The slice in the bottom panel is roughly at 3 o'clock in the upper left panel. Figures are provided courtesy of A. C. Boley.

the intense radiation field from the star and inner disk. It is even possible that some masers in massive protostars may be due to irradiation of the inner surfaces of spiral arcs (Durisen et al. 2001).

Boley & Durisen (2006) consider it highly likely that the "convective" motions reported by Boss (2004a) are actually dynamic vertical motions similar to those produced by hs-jumps. Boss (2004a) reports that his convective speeds are comparable to the sound speed and correlate with spiral arms, as expected for hs-jumps. Normal thermal convection is a quasistatic process occurring against an otherwise nearly hydrostatic background, and it

 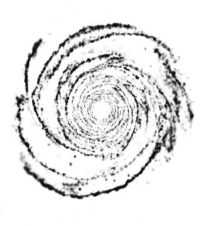

FIGURE 10. The concentration of 50 cm radius solid particles by gas drag in a GI-active disk. *Left.* The equatorial plane gas density structure for the $M_d/M_s = 0.25$ and $t_{cool}\Omega = 7.5$ disk of Lodato & Rice (2004) after about seven outer rotations. *Right.* The particle-to-gas density ratio for the disk in the left panel for 50 cm particles after the particles have interacted with the gas for only one outer rotation period. The darkest black represents an enhancement of the initial average particle to gas ratio by a factor of about 50. Figures adapted from Rice et al. (2004).

is difficult to imagine how an analog of convection can occur in a GI-active environment. Boss (2004a) does detect extensive superadiabatic gradients in the vertical direction, but these may be just another consequence of his different radiative boundary conditions. Some superadiabatic gradients are also seen in Mejía/Cai simulations, but it is not at all clear that they are related to thermal convection.

### 5.2. *Interaction of GIs with solids and contaminants*

As laid forth in classic papers by Weidenschilling (1977) and Völk et al. (1978), solid particles larger than dust grains experience net drifts in a disk relative to the gas. In a laminar disk with density and pressure decreasing radially and vertically outward, solids tend to orbit with Keplerian speeds, while the gas orbits somewhat more slowly due to pressure gradient forces. The resulting relative motions of the gas and solids induce drag forces that produce both radially inward and vertically downward drifts of the particles relative to the gas. Recently, Haghighipour & Boss (2003a,b) have explored the behavior of particles in a disk with a ring-like density maximum. Their ring, which is intended to represent a dense feature produced by GI activity, has a maximum at about 1 AU and a radial full width at half maximum that is also about an AU. Even for such a broad ring, solids migrate rapidly into the center. The particles with the shortest drift time scales, typically only a few hundred years, have sizes on the order of one meter.

The ring in Haghighipour & Boss (2003a,b) is a static structure. Rice et al. (2004) have injected solid particles into the $M_d/M_s = 0.25$ and $t_{cool}\Omega = 7.5$ simulation of a gravitationally unstable disk by Lodato & Rice (2004). The gas disk is allowed to establish asymptotic behavior before the introduction of the particles. Starting with an initially uniform particle distribution, the gas plus particle disk is then integrated for one more outer rotation, or about 125 years. The particle gravitational forces and the back reaction of drag forces on the gas are not included. Figure 10 illustrates the remarkably effective concentration of 50-cm-radius particles in this short amount of time. The strong

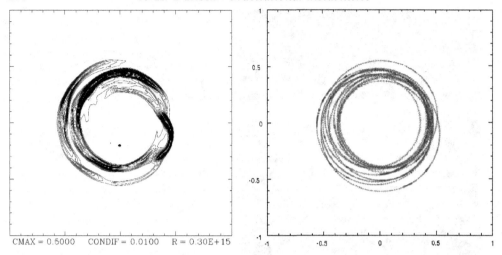

CMAX = 0.5000    CONDIF = 0.0100    R = 0.30E+15

FIGURE 11. *Left.* Midplane contours of contaminants initially painted on the surface of the disk over a narrow annulus at 9 AU show mixing to the midplane and radial spreading along spiral arms in only 120 years. *Right.* The line shows the trajectory of a multi-Jupiter mass accreting point mass inserted into a GI-active disk in place of a dense clump. In both panels the square is 40 AU along an edge. Figures adapted from Boss (2004b) and Boss (2005).

implication is that the marshalling of particles into dense structures by gas-dynamical GI activity can rapidly accelerate the formation of planetesimals. On the other hand, Rice et al. (2004) also find that an initially thin particulate disk is stirred to a finite thickness by the vertical motions associated with GIs.

Another question of interest is how a GI-active disk will affect the orbits of planet-sized objects, either those existing prior to the onset of GIs or formed by the GIs themselves. Although not subject to significant aerodynamic gas drag, planets interact with the surrounding gas through gravitational torques and migrate radially (Armitage, this volume). To study this effect, Boss (2005) assumes that dense clumps formed in his disk simulations are bound protoplanets and replaces them with accreting point masses. As shown in the right panel of Figure 11, an accreting point mass at about 10 AU has a fairly stable orbit radius despite the background of dense time-varying structure and net radial inflow in the disk due to GIs. In other words, protoplanets formed rapidly in a massive GI-active disk do not migrate rapidly inward. This is encouraging for the survival of gas giants, if they can indeed be formed promptly by GIs in a massive disk. Such orbital survival of at least some protoplanets formed directly by GIs was already demonstrated by the SPH simulations of Mayer et al. (2004).

In addition to considering large condensed objects, Boss (2004b) also addresses the question of how contaminants, abundance anomalies, and small entrained particles might be mixed by GI turbulence. In a series of experiments, he creates a narrow region of altered composition and solves a 3D continuity plus diffusion equation for this "color" contaminant simultaneously with the hydrodynamics equations. One such experiment, seen in the left panel of Figure 11, shows that within only 120 years, an anomalous abundance painted at high disk altitude over a 1 AU-wide annulus at 9 AU has propagated to the midplane and spread radially. The contaminants tend to trace the dense spiral arms of the GIs, as would be expected if the mixing flows are hs-jumps (Fig. 9).

So, work to date indicates that GIs are likely to accelerate planetesimal formation by concentrating moderate-sized solids into dense structures, but they do not cause wholesale inward migration of large embedded solids. GIs are also effective at mixing fine-grained and gaseous components of the disk.

### 5.3. *Combined effect of GIs and MRIs*

It will eventually be necessary to understand how GIs interact with the presence of other disk transport mechanisms. The magnetorotational instabilities (MRIs), which occur in any disk with sufficient conductivity, are a likely source of turbulence (Balbus & Hawley 1998). It is possible for GIs and MRIs to coexist in disks in separate or even overlapping regions. Gammie (1996) points out that, for disks around young solar-type stars, there could be a region near the midplane beyond the innermost disk, called the "dead" zone, where MRIs do not occur due to the low degree of ionization of this cool and shielded gas. Accretion could continue by MRI turbulence in thin layers near the disk surface, but material in the dead zone would otherwise tend to accumulate until GIs set in. Using a simple $\alpha$ prescription for both MRIs and GIs, Armitage et al. (2001) model the behavior of a disk with a dead zone and find that, for moderate average infall rates onto the disk of about $10^{-6}$ $M_\odot$/yr, the disk exhibits episodic accretion outbursts at intervals of $10^4$ to $10^5$ yrs, with peaks inflow rates reaching $10^{-5}$ $M_\odot$/yr. The peak accretion rate is similar to the inflow rates seen during the burst phase of $t_{\rm cool}$ = constant simulations (§3.1).

Although the Armitage et al. calculations are suggestive, the adequacy of an $\alpha$ treatment for the behavior of GIs is somewhat questionable (§3.3). A step in the direction of a full blown global treatment has been taken in Fromang et al. (2004), where magnetized and self-gravitating adiabatic disks are evolved in 3D using a magnetohydrodynamics (MHD) code. When both mechanisms operate in their disks, they find significant interactions of global modes which tend to damp the GIs by tens of percents and also produce dynamic oscillations between low and high values of gravitational stress. In addition to strong global modes, the disk becomes suffused with turbulence. Only a limited parameter range is explored so far by Fromang et al. (2004), but there is potential for a rich phenomenology when both processes operate in the same disk. More research along these lines is necessary.

### 5.4. *Hybrid planet formation scenario*

Although there may be various ways to make core accretion plus gas capture work faster (Hubickyj, this volume) than in the classic models of Pollack et al. (1996), indications of planet formation at very early times around young stars like CoKu Tau/4 (Quillen et al. 2004) remain challenging for core accretion. When type I migration of growing cores is ignored, probabilistic treatments based on planet growth and disk evolution modeling can be used to fit the correlation of planet detection with central star metallicity in the framework of core accretion (Ida & Lin 2004; Kornet et al. 2005), but rapid type I migration of the cores inward into the star remains a vexing problem (Armitage, this volume).

As discussed in Section 4, it is not clear from currently available simulations that disks ever cool fast enough for direct planet formation by fragmentation; but there is significant agreement (e.g., Boss & Durisen 2005) that GIs will produce dense structures in disks, including spiral arms and arcs. Some of the most long-lived, well-defined features seen in nonfragmenting simulations, with $t_{\rm cool}$ = constant (Pickett et al. 2003; Mejía et al. 2005a) and with the Mejía/Cai treatment of radiative cooling (Mejía et al. 2005b; Cai et al. 2006), are dense gas rings near the boundary between a hot inner disk and an outer GI-active disk. An analysis by Durisen et al. (2005) argues that ring formation is expected to occur near boundaries between GI-active and inactive regions. In Figure 12, ring growth appears to be due to a pattern of overlapping Lindblad resonances in the GI-

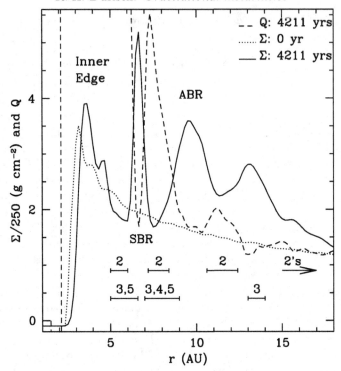

FIGURE 12. Inner rings of a $t_{cool} = 2$ simulation at a late time in the asymptotic phase. The $\Sigma$ curves are azimuthally averaged surface densities. The ABR and SBR, the *active* and *secondary* boundary rings, are nearly axisymmetric in structure. The numbered bars along the bottom indicate where inner Lindblad resonances with outer disk patterns can be detected by Fourier analysis. The numbers indicate the $m$-value or armedness of the modes. Note that resonances seem to straddle all strong radial concentrations of mass, including the ABR and SBR. This figure is from Durisen et al. (2005).

inactive disk interior to 10 AU produced by GIs in the active region beyond 10 AU. As in Fromang et al. (2004), alignments of corotation and Lindblad resonances seem to play a role in producing the radial hierarchy of modes. Note that the surface density of the ring labeled SBR, which contains about six Jupiter masses by the end of the calculation, has almost tripled in only 4,000 years. This simulation also suggests that waves propagate into the GI-inactive region from the GI-active side. Dissipation of these waves may keep the inner disk hot and may be a manifestation of the energy transport by waves expected for global GI behavior (Balbus & Papaloizou 1999).

   The formation of rings and other dense structures by GIs, combined with the tendency for gas drag to sweep meter-sized and larger particles into them, as demonstrated by Haghighipour & Boss (2003a) and Rice et al. (2004), lead Durisen et al. (2005) to propose a hybrid planet-formation scenario. Long-lived, ring-like structures produced by GIs provide a safe haven in which cores can grow rapidly as solids drift to the ring centers, thereby accelerating planet formation by core accretion plus gas capture. Similar suggestions, that solids might accumulate at the centers of stable vortices or at disk edges, have been made previously by Klahr & Henning (1997) and Bryden et al. (2000), respectively. Dense gas rings additionally act as a barrier to type I migration and an inhibition to gap formation, and so the cores can grow to reasonable sizes and accrete the large envelopes necessary to make super-Jupiter mass objects. Self-gravity of the gas in the rings may also shorten the time necessary to accrete an envelope. Simulations to confirm the reality and robustness of GI-induced ring formation are underway. Core accretion simulations in the environment of a ring would also be very useful.

## 6. Conclusions

### 6.1. *The big questions revisited*

Returning to the big questions presented at the beginning of this review, we can see that some of them are addressed head on by recent simulations, and the answers to others can be inferred.

#### 6.1.1. *Do gravitational instabilities produce gas giant planets?*

Given the formation of dense gas structures by GIs and the tendency of solid particles to concentrate into these features, it seems likely that GIs play a significant role in planet building.

*Do they form planets directly on a dynamic time scale?* Probably not. The answer to this question is controversial at several levels. While all current researchers agree that, with rapid enough cooling, growing GI spiral waves in disks will fragment into dense clumps, there are disagreements about whether or not such clumps become bound protoplanets (§2.3). More seriously, there are sharp disagreements about whether radiative or convective cooling in real disks is ever rapid enough for fragmentation to occur in the first place (§4.2). Recent results (§4.3) by my own hydro group suggest that environmental factors, especially irradiation, can have a damping effect on GIs and that GI behavior may depend more on metallicity and grain size than indicated by work published to date. Although it is premature to rule out direct gas giant formation by GIs on a dynamic time scale, I think the weight of evidence right now tilts away from that conclusion. Realize, however, that we are far from probing the full complexity of this problem. There are regimes yet unexplored, such as behaviors at major opacity boundaries (Johnson & Gammie 2003), equation of state effects, and fully coupled evolution of solids and gas.

*Do they accelerate core accretion, instead?* Probably. The formation of long-lived rings in some GI simulations, especially at the boundaries between GI active and inactive regions (§3.1 and §5.4), and the rapidity with which meter-sized solids can migrate to the centers of these and other GI structures (§5.2) opens the possibility that, instead of being an alternative gas giant formation mechanism to core accretion plus gas capture, GIs provide the environment in which core accretion can be accelerated to very short times without rapid loss of growing cores into the star due to type I migration. The calculations by Boss (2005), while presented as support for direct formation by disk instability, also show that massive bodies can maintain relatively stable orbits against the background of a GI-active disk. Rings, however, probably provide the most uniform and nurturing milieu for accelerated core accretion. It is important for other researchers to confirm or refute their occurrence, understand the mechanisms that form and sustain them, and determine their ultimate fate.

### 6.1.2. *How fast can GIs transport mass in a disk?*

Simulations of GIs in Solar System-sized disks generally have inner regions that are not directly computed, and so accretion induced by GIs cannot be followed all the way down to the star. However, global 3D calculations show that gravitational torques produce mass transport, with inflow predominating over the inner and middle disk regions. Angular momentum removed from the inner disk is transported to the outer disk, which expands. The mass accretion rates in the inflow regions are typically reported to be in the range

$10^{-7}$ to $10^{-5}$ $M_\odot$ yr$^{-1}$ (Nelson et al. 2000; Pickett et al. 2003; Boss 2002b; Lodato & Rice 2004; Mejía et al. 2005a,b), which would be healthy mass inflow rates for a young circumstellar gas disk. With a disk mass of 0.1 $M_\odot$, these rates lead to disk lifetimes between $10^4$ to $10^6$ years. In the longest simulations to date, GIs appear to settle into an asymptotic behavior where mass accretion is sustained at rates in the lower end of the range for at least many thousands of years and probably much longer (§3.1). The inflow process is variable on a dynamic time scale, with large fluctuations about the average.

Inflow values near $10^{-5}$ $M_\odot$ yr$^{-1}$ are attained over shorter time scales as GIs initiate in simulations with a global $t_{\rm cool}$ (§3.1 and §4.3) or during the eruptions of a dead zone (§5.3). Lodato & Rice (2005) find peaks as high as $10^{-4}$ $M_\odot$ yr$^{-1}$ in the outer disk for $M_d = M_s$. Boss (2002b) even reports a mass inflow rate of $10^{-3}$ $M_\odot$ yr$^{-1}$. Further modeling of inner disk regions and of dead zone behaviors are needed to determine whether peak accretion rates onto the star are high enough to fuel an FU Orionis outburst.

*Is this process local or global?* If GIs behave in a local manner, where mass and energy transport can be described accurately using only local disk properties, then they are probably susceptible to an α-type prescription, which is desirable for evolving disks with GIs over long time intervals (§3.2). Simulations suggest that "locality" depends on several factors. First of all, if disks are extremely massive ($M_d/M_s \geqslant 0.25$) and thick (scale height $> 0.1r$), then GIs will behave globally in some respects regardless of other constraints. For less massive disks with moderate to small scale heights, the "locality" of GIs depends on whether the cooling time $t_{\rm cool}$ behaves locally (i.e., $t_{\rm cool}\Omega \approx$ constant) or globally (i.e., $t_{\rm cool} \approx$ constant; §3.3). For local $t_{\rm cool}$s, equation (3.1) appears to be accurate; for global cases, the mass inflow greatly exceeds what is expected from (3.1). Simulations of disks with radiative cooling (§4.3) show that real disks are probably better described as having global $t_{\rm cool}$s. Gravitational stresses dominate in all 3D global simulations, but are matched by Reynolds stresses in thin-disk local simulations. When GIs are global, transport is mediated by a few low-order, usually two-armed, modes. A key point that remains to be resolved is whether nonlocal energy transport by waves occurs in global $t_{\rm cool}$ calculations.

*Does it produce persistent structures, like dense arms and rings?* Yes. Simulations integrated for tens of orbits indicate relaxation of disks into an asymptotic behavior with persistent turbulent spiral wave structure. In simulations where GIs behave globally, long-lived dense rings may also grow near boundaries between GI-active and inactive parts of the disk.

### 6.1.3. *When do GIs occur in disks?*

GIs may occur during the protostellar core collapse phase as a massive disk first forms (e.g., Laughlin & Bodenheimer 1994) and at later phases of disk evolution due to mass accumulation in a dead zone at a few AU from the star (e.g., Gammie 1996; Armitage et al. 2001) or in outer disk regions due to a fall off in the value of $Q$ with increasing $r$ (e.g., D'Alessio et al. 1999).

*Do they occur in the early embedded phase?* It depends. Although it is commonly assumed that this is the prime time to have vigorous GIs in protoplanetary disks, the results discussed in Section 4.3.1 sound a cautionary note. Irradiation, if sufficiently strong, may suppress GIs by preventing the disk from reaching unstable values of $Q$. The same heating also tends to thicken the disk and result in greater stability than given by the usual $Q$-criterion (see Mayer et al. 2004). The message is that one has to consider the detailed structure of a disk and the intensity of its radiation environment in order to properly assess its stability and the strength of any resultant GIs.

*Do they occur in dead zones?* Probably. A thorough discussion of ionization conditions in protoplanetary disks goes beyond the scope of this review, but dead zones (§5.3) of low ionization, where MRIs cannot operate, are likely to exist and may lead to episodic bursting GI behavior (e.g., Armitage et al. 2001). Further modeling of this situation in full 3D with radiative hydrodynamics could reveal connections with FU Ori outbursts, thermal processing of solids, and planet formation (Boss & Durisen 2005a). To understand this fully, magnetic fields (Fromang et al. 2004) have to be included.

### 6.1.4. *How do GIs affect solids and contaminants?*

This could be the most exciting area of current development in the study of GIs in disks.

*Do they mix solids and contaminants?* Yes. Studies by Boss (2004b) and Boley et al. (2005; SS5.1 and 5.2) show that motions associated with the turbulent spiral structure of a GI-active disk can mix contaminants and entrained dust grains both vertically and horizontally over tenths of AU scales on a dynamic time scale. This mixing tends to correlate with the dense spiral wave structure of the disk.

*Do they concentrate solids into coherent structures?* Yes. Haghighipour and Boss (2003a,b) and Rice et al. (2004) demonstrate extremely short timescales for meter-sized solids to be swept into both relatively static ring-like structures and dynamic GI spiral waves. It is very likely then that planetesimal formation and perhaps core growth can be accelerated by GIs.

*Do they cause thermal processing of solids?* Yes. Boss & Durisen (2005a) and Boley et al. (2005) show that the dense clumps and spirals in GI-active disks can produce strong enough shocks to process solid materials.

### 6.2. *Parting thoughts*

The study of gravitational instabilities in disks and of their relationships to planet formation and disk evolution is on the verge of becoming a mature research area, where a full array of relevant processes are treated by a variety of sophisticated methods and where consensus builds on a broad foundation of key results and principles. Neither criterion for maturity currently applies, but pathways are emerging along which we can make progress toward this goal. Fundamental disagreements remain about cooling times for disks and longevity of fragments. These will hopefully be resolved through the participation of more research groups and through collaborations by existing groups for a common purpose, such as the one now being led by Lucio Mayer on fragmentation (2005, private communication). At the same time, there are a few consensus results in the bag, such as the Gammie fragmentation criterion and the central importance of thermal physics to GI behavior.

At the moment, I find it most exciting to have some tantalizing initial tastes of the deliciously complex interplay that must occur between solids and gas through both opacity effects and dynamics. This could eventually engender a significant paradigm shift in our view of planet formation towards scenarios involving *both* GIs *and* core accretion plus gas capture. Similar appetizers are offered by efforts to blend the two most important disk transport mechanisms, GIs and magnetorotational instabilities, into a coherent global picture. I invite more researchers to break out their formal wear (or software) and join the coming banquet.

I would like to thank P. J. Armitage, A. C. Boley, A. P. Boss, K. Cai, D. N. C. Lin, J. J. Lissauer, G. Lodato, L. Mayer, A. C. Mejía, S. Michael, M. K. Pickett, J. E. Pringle, and W. K. Rice for useful recent discussions or email and, in some cases, for specific help with this manuscript. Extra special thanks are due to A. C. Boley for dealing with troublesome graphics format conversions. I was supported during this work by NASA Origins of Solar Systems grant NAG5-11964.

## REFERENCES

ARMITAGE, P. J., LIVIO, M., & PRINGLE, J. E. 2001 *MNRAS* **324**, 705.

BALBUS, S. A. & HAWLEY, J. 1999 *Rev. Mod. Phys.* **70**, 1.

BALBUS, S. A. & PAPALOIZOU, J. C. B. 1999 *ApJ* **521**, 650.

BATE, M. R. & BURKERT, A. 1997 *MNRAS* **228**, 1060.

BELL, K. R., CASSEN, P. M., WASSON, J. T., & WOOLUM, D. S. 2000. In *Protostars and Planets IV* (eds. V. Mannings, A. P. Boss, & S. S. Russell). p. 897. Univ. Arizona Press.

BODENHEIMER, P., YORKE, H. W., RÓZYCZKA, M., & TOHLINE, J. E. 1990 *ApJ* **355**, 651.

BOLEY, A. C. & DURISEN, R. H. 2005 *ApJ*, **641**, 534.

BOLEY, A. C., DURISEN, R. H., & PICKETT, M. K. 2005. In *Chondrites in the Protoplanetary Disk* (eds. A. N. Krot, E. R. D. Scott, & B. Reipurth). ASP Conf. Ser. 341, p. 839. ASP.

BOSS, A. P. 1997 *Science* **276**, 1836.

BOSS, A. P. 1998 *ApJ* **503**, 923.

BOSS, A. P. 2000 *ApJ* **536**, L101.

BOSS, A. P. 2001 *ApJ* **563**, 367.

BOSS, A. P. 2002a *ApJ* **567**, L149.

BOSS, A. P. 2002b *ApJ* **576**, 462.

BOSS, A. P. 2003 *LPI* **34**, 1075.

BOSS, A. P. 2004a *ApJ* **610**, 456.

BOSS, A. P. 2004b *ApJ* **616**, 1265.

BOSS, A. P. 2005 *ApJ* **629**, 535.

BOSS, A. P. & DURISEN, R. H. 2005a *ApJ* **621**, L137.

BOSS, A. P. & DURISEN, R. H. 2005b. In *Chondrites in the Protoplanetary Disk* (eds. A. N. Krot, E. R. D. Scott, & B. Reipurth). ASP Conf. Ser. 341, p. 821. ASP.

BRYDEN, G., RÓZYCZKA, M., LIN, D. N. C., & BODENHEIMER, P. 2000 *ApJ* **540**, 1091.

CAI, K., DURISEN, R. H., MICHAEL, S., BOLEY, A. C., MEJÍA, A. C., PICKETT, M. K., & D'ALESSIO, P. 2006 *ApJ*, **636**, L149; Erratum—ibid. 2006 **642**, L173.

CALVET, N., PATINO, A., MAGRIS, G. C., & D'ALESSIO, P. 1991 *ApJ* **380**, 617.

CAMERON, A. G. W. 1978 *Moon & Planets* **18**, 5.

CHIANG, E. I. & GOLDREICH, P. 1997 *ApJ* **490**, 368.

CHICK, K. M. & CASSEN, P. 1997 *ApJ* **477**, 398.

D'ALESSIO, P., CALVET, N., & HARTMANN, L. 2001 *ApJ* **553**, 321.

D'ALESSIO, P., CALVET, N., HARTMANN, L., LIZANO, S., & CANTO, J. 1999 *ApJ* **527**, 893.

D'ALESSIO, P., CANTO, J., CALVET, N., & LIZANO, S. 1998 *ApJ* **500**, 411.

DESCH, S. J. & CONNOLLY, H. C., JR. 2002 *Meteor. Planet. Sci.* **37**, 183.

DURISEN, R. H. 2001. In *The Formation of Binary Stars* (eds. H. Zinnecker & R. D. Mathieu). IAU Conf. Ser. 200, p. 381. ASP.

DURISEN, R. H., CAI, K., MEJÍA, A. C., & PICKETT, M. K. 2005 *Icarus* **173**, 417.

DURISEN, R. H., GINGOLD, R. A., TOHLINE, J. E., & BOSS. A. P. 1986 *ApJ* **305**, 281.

DURISEN, R. H., MEJÍA, A. C., & PICKETT, B. K. 2003 *Rec. Devel. Astrophys.* **1**, 173.

DURISEN, R. H., MEJÍA, A. C., PICKETT, B. K., & HARTQUIST, T. W. 2001 *ApJ* **563**, L157.

FISCHER, D. A. & VALENTI, J. 2005 *ApJ* **622**, 1102.

FROMANG, S., BALBUS, S. A., TERQUEM, C., & DE VILLIERS, J.-P. 2004 *ApJ* **616**, 364.

GAMMIE, C. F. 1996 *ApJ* **457**, 355.

GAMMIE, C. F. 2001 *ApJ* **553**, 174.

GOLDREICH, P. & LYNDEN-BELL, D. 1965 *MNRAS* **130**, 125.

HAGHIGHIPOUR, N. & BOSS, A. P. 2003a *ApJ* **583**, 996.

HAGHIGHIPOUR, N. & BOSS, A. P. 2003b *ApJ* **598**, 1301.

HEINEMANN, T., DOBLER, W., NORDLUND, Å, & BRANDENBURG, A. 2006 *A&A*, **448**, 731.

IDA, S. & LIN, D. N. C. 2004 *ApJ* **616**, 567.

JOHNSON, B. M. & GAMMIE, C. F. 2003 *ApJ* **597**, 131.

JOHNSTONE, D., HOLLENBACH, D., & BALLY, J. 1998 *ApJ* **499**, 758.

KLAHR, H. H. & HENNING, T. 1997 *Icarus* **128**, 213.

KORNET, K., BODENHEIMER, P., RÓŻYCZKA, M., & STEPINSKI, T. F. 2005 *A&A* **430**, 1133.

KUIPER, G. P. 2001. In *Proceedings of a Topical Symposium* (ed. J. A. Hynek). p. 357. McGraw-Hill.

LAUGHLIN, G. & BODENHEIMER, P. 1994 *ApJ* **436**, 335.

LAUGHLIN, G., KORCHAGIN, V., & ADAMS, F. C. 1997 *ApJ* **477**, 410.

LAUGHLIN, G., KORCHAGIN, V., & ADAMS, F. C. 1998 *ApJ* **504**, 945.

LAUGHLIN, G. & RÓŻYCZKA, M. 1996 *ApJ* **456**, 279.

LIN, D. N. C. & PRINGLE, J. E. 1987 *MNRAS* **225**, 607.

LODATO, G. & RICE, W. K. M. 2004 *MNRAS* **351**, 630.

LODATO, G. & RICE, W. K. M. 2005 *MNRAS* **358**, 1489.

LUBOW, S. H. & OGILVIE, G. I. 1998 *ApJ* **504**, 983.

MARTOS, M. A. & COX, D. P. 1998 *ApJ* **509**, 703.

MATZNER, C. D. & LEVIN, Y. 2005 *ApJ* **628**, 817.

MAYER, L., QUINN, T., WADSLEY, J., & STADEL, J. 2002 *Science* **298**, 1756.

MAYER, L., QUINN, T., WADSLEY, J., & STADEL, J. 2004 *ApJ* **609**, 1045.

MEJÍA, A. C. 2004 *Ph.D. dissertation*, Indiana University.

MEJÍA, A. C., DURISEN, R. H., PICKETT, M. K., & CAI, K. 2005 *ApJ* **619**, 1098.

MEJÍA, A. C., DURISEN, R. H., PICKETT, M. K., CAI, K., & D'ALESSIO, P. 2005 *ApJ*, **619**, 1098.

NELSON, A. F. 2003. In *Scientific Frontiers in Research on Extrasolar Planets* (eds. D. Deming & S. Seager). ASP Conference Series, p. 291. ASP.

NELSON, A. F., BENZ, W., ADAMS, F. C., & ARNETT, D. 1998 *ApJ* **502**, 342.

NELSON, A. F., BENZ, W., & RUZMAIKINA, T. V. 2000 *ApJ* **529**, 1034.

OSORIO, M., D'ALESSIO, P., MUZEROLLE, J., CALVET, N., & HARTMANN, L. 2003 *ApJ* **586**, 1148.

PACZYŃSKI, B. 1978 *Acta Astron.* **28**, 91.

PAPALOIZOU, J. C. B. & SAVONIJE, G. 1991 *MNRAS* **248**, 353.

PICKETT, B. K., CASSEN, P., DURISEN, R. H., & LINK, R. 1998 *ApJ* **504**, 468.

PICKETT, B. K., CASSEN, P., DURISEN, R. H., & LINK, R. 2000 *ApJ* **529**, 1034.

PICKETT, B. K., DURISEN, R. H., & DAVIS, G. A. 1996 *ApJ* **458**, 714.

PICKETT, B. K., MEJÍA, A. C., DURISEN, R. H., CASSEN, P. M., BERRY, D. K., & LINK, R. P. 2003 *ApJ* **590**, 1060.

POLLACK, J. B., HOLLENBACH, D., BECKWITH, S., SIMONELLI, D. P., ROUSH, T., & FONG, W. 1994 *ApJ* **421**, 615.

POLLACK, J. B., HUBICKYJ, O., BODENHEIMER, P., LISSAUER, J. J., PODOLAK, M., & GREENZWEIG, Y. 1996 *Icarus* **124**, 62.

QUILLEN, A. C., BLACKMAN, E. G., FRANK, A., & VARNIÈRE, P. 2004 *ApJ* **612**, L137.

RAFIKOV, R. R. 2005 *ApJ* **621**, L69.

RICE, W. K. M., ARMITAGE, P. J., BATE, M. R., & BONNELL, I. A. 2003 *MNRAS* **339**, 1025.

RICE, W. K. M., LODATO, G., PRINGLE, J. E., ARMITAGE, P. J., & BONNELL, I. A. 2004 *MNRAS* **355**, 543.

TOOMRE, A. 1964 *ApJ* **139**, 1217.

TOMLEY, L., CASSEN, P., & STEIMAN-CAMERON, T. Y. 1991 *ApJ* **382**, 530.

TRUELOVE, J. K., KLEIN, R. I., MCKEE, C. F., HOLLIMAN, J. H., II, LOWELL, L. H., & GREENOUGH, J. A. 1997 *ApJ* **489**, L179.

VÖLK, H. J., MORFILL, G. E., RÖSER, S., & JONES, F. C. 1978 *Moon & Planets* **19**, 221.

WEIDENSCHILLING, S. J. 1977 *MNRAS* **180**, 57.

WOOD, J. A. 1996 *Meteor. Planet. Sci.* **31**, 641.

YORKE, H. W. & BODENHEIMER, P. 1999 *ApJ* **525**, 330.

YOUDIN, A. N. & SHU, F. H. 2002 *ApJ* **580**, 494.

# Conference summary: The quest for new worlds

## By J. E. PRINGLE

Institute of Astronomy, Madingley Road, Cambridge, CB3 0HA, UK

## 1. Introduction

I would first of all like to express my gratitude to the organizers of this meeting for inviting me to give this summary talk. I am assured by my colleagues that such an invitation is one of the key prerequisites for becoming recognized as an "Old Fart"—so, "Thank you, Mario." To reinforce this point, I shall be the first speaker to demonstrate that the overhead projector is still working, thus blocking the view of those who like to sit in the middle of the front row. Although I am not (yet) one of those who feels it necessary to demonstrate status by only attending a part of a meeting, I must admit that this is the first meeting I have been to in which I have attended every talk—I use the word "attended" deliberately, since initially the effects of jet lag had not quite worn off.

I take, however, neither credit nor responsibility, for the slightly pretentious title, *The Quest for New Worlds*. As I am sure you are all aware, this was the title of the grant proposal to the Spanish authorities from a certain 15th century Italian. He set out to find India, but instead discovered the local *bête noire*, Cuba. For this, he is celebrated here each year in October, though one must admit a certain admiration for someone who has the ability to keep getting grant proposals funded, despite a complete failure to achieve the stated goals and objectives.

Before taking a closer look at what we have learned over the last few days—and what we didn't—it is instructive to step back and take a broader look at the field as a whole. To do so let me share with you a couple of quotes.

The first is from a colleague, who is an inveterate and assiduous attender of conferences, who was being questioned about latest set of general conferences (Texas Symposium, AAS Meeting, ...) he had just come back from. When asked what was new in planets, the response was to the effect that it is not a field in which much seems to be happening, and that the field itself is rather slow moving. Thus, Steve Beckwith's comments in his introductory talk that we are dealing with the fundamental question "Are we alone?"—that this area serves as one of the drivers of the Philosophy of Science, and that we find ourselves in the borderline between Science and Religion—do not appear to have reached all of our colleagues. Mind you, our cosmological colleagues, with their "fingers of God," and the *COBE* results showing us the face of God, already seem to be developing their own theological terminology. In this context, it is worth recalling that Immanuel Kant, who we now recognize in his treatise *Allgemeine Naturgeschichte und Theorie des Himmels* as having been a pioneer in the theory of the formation of planetary systems, devoted much care and tact to reassuring his sponsors, as well as the powers that be, that he was not casting doubt on Who created the Heavens, but was, rather, humbly considering how He might have set about it.

The second is from a colleague in the field of a theoretical bent. The colleague had just some back from some recent meetings in the field and was being asked to report.

The general tone of the response was that the field is "getting a bit boring—there are far too many observations." As Steve Beckwith remarked in his introduction, this field is predominantly an empirical one, and much of what has been found has defied ready predictability. From my notes I see that he also said something about "idle speculation," which is probably a reference to the efforts of we theorists. In any case, it is certainly one of those areas of astrophysics in which one hears the merry scampering of theorists trying to keep up. From a theorist's perspective, as the observations pile in, a field can move from one of excitement, in which one receives continual stimulation and feels that one has a chance of explaining something, to one of despair, in which one starts to wonder if one can ever explain anything. I think we would all agree that this field is just building up to the excitement stage.

Indeed a more objective test of the vitality of a field is to ask: Would you recommend a new graduate student to go into it? I started my PhD just after the launch of the *UHURU* X-ray Telescope. There was a wealth of new observational data, which led to a certain buzz about the field. Some 35 years later that field is still booming, led predominantly by continuing advances in instrumental capabilities. I get that same feeling here—this is an exciting field, still growing in its observational capabilities, and the best is yet to come. Like the early days of x-ray astronomy, it is a young person's field. It is definitely a good one to start a career in. Indeed, for myself it has been sobering to see that the launch dates for the spacecraft being discussed at this meeting are close to, and some beyond, my retirement date at the University of Cambridge.

## 2. A field with a past

However, while the field of the formation of planetary systems is a young one, it is worth bearing in mind that, unlike x-ray astronomy in 1970, it is a field with a history. The point here is that just over 10 years ago, the field of planet formation was in really good shape. At that time, theorists were able to predict that if one starts with a disk of gas about 30 AU in radius, with a mass of around 0.01 solar masses, around a star of about one solar mass, then the inevitable outcome is the formation of a Jupiter-mass planet at around 5 AU, a Saturn-mass planet at around 10 AU, an Earth sized, terrestrial-type planet at around 1 AU, and so on. The success of this picture has had a very strong influence on our thinking, and it is good to retain an awareness of this fact, in case our preconceptions start to lead us astray. To give an example, some 15 years ago, Doug Lin and I wrote a paper on the "Initial evolution of protostellar disks." Our disks started life as massive (a good fraction of the mass of the central star) and large (a few hundred to a thousand AU). We got ridiculed by some of the pundits at that time, because, we were told, it was clear from the Solar System that disks are neither that massive, nor that big. Thus, when I hear talk of "hot Neptunes," without any evidence that the objects in question are anything like Neptune, I start to worry a little. And to hear one speaker talking glibly about 'other solar systems' sets alarm bells ringing. The concern here is the extent to which perversion of the language can lead on to muddled thinking.

## 3. A field with a future

Over the last 10 years there has been a steady increase in the number of planetary systems that we know about. Indeed, as we have heard, there are already over 100 planetary systems known. One thing is clear from what we have found so far, and this is that none of the systems found so far is similar to the Solar System. In Figure 1, I show an updated version of the plot given in Beer et al. (2004). This shows eccentricity plotted

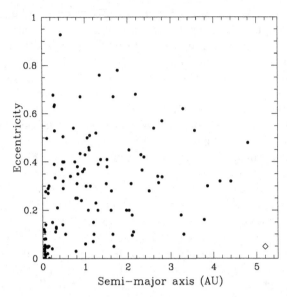

FIGURE 1. A plot of eccentricity versus semi-major axis of the planet that induces the largest velocity semi-amplitude in each of the observed systems. The diamond at the bottom right represents the Solar System in the form of Jupiter.

against periastron for the largest contributor to the radial velocity detection for each of the known planetary systems. Also shown is the point for Jupiter's orbit to represent the Solar System. It is quite evident that in this plot the Solar System is an outlier. None of the planetary systems discovered so far have their major planet in the right place or in the right orbit. Compared to the Solar System, the planetary systems found so far have their planets significantly closer to the central star, and with significantly higher orbital eccentricity (except for those so close in that tidal circularization has taken place). In terms of some form of the "Anthropic Principle" (summed up by the words of the First World War song "We're here because we're here because we're here because we're here"), one could make a *prima facie* case that the Solar System is special in some way. But I think it is fair to say that at the moment there is no one who really believes that, when the Domesday Book of planetary systems has been completed, the Solar System will look out of place. However, in science, belief is not enough, and what this implies, therefore, is a general belief that our roll call of planetary systems is seriously incomplete.

We have heard a number of talks at this meeting about the continued search for new planetary systems. It is clear that there are strong observational biases which explain why the planetary systems discovered so far have been found at small radius, because for those systems the radial velocities are higher and the orbits shorter. Waiting for longer orbital periods to be found may just be a matter of time, but it is clear from the talks by both Marcy and Mayor that we are approaching the limitations of the radial velocity technique, not least because of intrinsic stellar properties such as pulsations. In this context, Doug Lin put in a plea that some effort be made to see what constraints can be set by the large number of null, or near null, results already obtained, and the talk by Bob Brown gave some suggestions as to how progress might be made in this area. However, in order to extend the parameter space of planetary systems, it is evident that other techniques will be required. Of these, the direct techniques involving transits will again find mainly very short-period planets, whereas the micro-lensing searches will tend to find planets at intermediate separations of around a few AU. So it may be some time before the picture can be much extended, let alone completed.

## 4. The field now

That, for the time being, is for the future. At present, the evidence we have is that the planetary systems found so far have their planets in the wrong place and in the wrong orbits. If you are a theorist and are putting together a model or explanation for some phenomenon, then you feel that you are doing pretty well if you are able to account for more than around 90–95% percent of the observational "facts." Actually, some of us feel we are doing well if we can manage half of them. With this in mind, there have been two main approaches to developing theories of planetary formation. The first, and the most popular, is to follow the advice in Hilaire Belloc's poem, which is "Always keep a-hold of nurse, for fear of finding something worse." The second, is to take the *Monty Python* refrain "And now for something completely different." Let's look at the second first.

### 4.1. *The* Monty Python *approach*

The main alternative strategy is to try to form planets directly, using gravity alone. It seems to work for stars, so why not for planets? Using this approach one would expect to find only gas giants—thus no terrestrial planets and not much astrobiology. But, except for the Solar System, this would not contradict what we currently know. For example, there is little evidence as yet that extra-solar planets have cores.

The current status of this approach was well summarized in the talk by Richard Durisen. The point here is that since stars form with angular momentum, it seems likely that most of the mass of a star is, at some stage, processed through the disk. Thus, in the early stages at least, the disks should be self-gravitating. Most of the computational simulations of this process start with a sufficiently massive disk around a central point mass that it is self-gravitating and then let go. This is, of course, not realistic, but numerically probably the best one can do. Some authors find that planet-sized objects condense out, and some do not. But the general consensus at present appears to be that the generation of long-lived planetary mass objects has yet to be demonstrated. The major problem appears to be that, in order to get self-gravity to produce individual objects, rather than just produce spiral arms which accelerate the accretion process, it is necessary for the gas to be able to cool rapidly. Thus, good, accurate treatment of radiative processes will be required. There are also numerical issues. Modeling the evolution of compact structures in an accretion disk calls out for the use of a Lagrangian numerical method such as SPH—grid methods are not well suited to following compact structures moving supersonically across the grid. However, radiative transfer is more easily accurately done in a grid environment.

The basic conclusion seems to be that so far the viability of this approach is "not proven." The signs at present do not look encouraging, but there are still a lot of possibilities to explore.

### 4.2. *Keeping a-hold of nurse*

As we have heard at this meeting, most work on the formation of planetary systems involves keeping "a-hold of nurse." That is, the favored strategy is to take the original set of ideas for the formation of the Solar System and set about modifying them to fit the extra-solar data. A glance through the various review talks presented here shows that there has been a lot of progress. Understandably, most speakers have emphasized what works. I'll focus here a bit more on what doesn't work, and what needs to be done next.

There are basically two sets of problems here—problems with the Solar System story, and problems with the modifications required.

The general narrative of planet formation in the Solar System contains a number of miracle moments. The general aim is to start with a gaseous disk of material, allow the heavier elements to condense out as dust, to allow the dust to settle and coagulate into ever larger entities going from pebbles to rocks to planetesimals, and thence to planetary cores. The details of many of these processes still remain obscure. Having formed planetary cores, it is then necessary to persuade the gas to accrete onto it (to form gas giants) fast enough, before the disk has dispersed. This process was discussed in the talks by Calvet, Lissauer and Hubickyj. Another problem here, mentioned briefly by Armitage, is the one of Type I migration. This is the regime when a small planet, or protoplanet, has formed in the disk and is not massive enough to open up a gap in the surface density. Although the estimates of the migration timescales (both analytic and numerical) for this process have been steadily increasing, the current values of the migration timescales are still too short, by about an order of magnitude, to fit comfortably with observations.

The main modifications which need to be made to this standard picture are two-fold. First, we require radial migration in order to form the gas giants in the standard place and then move them inwards to where they are found. Second, we need to generate the observed eccentricities.

In general, as shown in the talk by Phil Armitage, and commented upon by Marcy, the migration story seem to work fairly well (as long as one can somehow suppress Type I migration) and the radial distribution of the currently observed systems can be reasonably accounted for. The more serious problem seems to be the one of generating the required distribution of eccentricities. The basic problem here is that accretion of gas to form gas giants and radial migration both require gas in the disk, and a planetary orbit in a gaseous disk that becomes and remains close to circular. And yet, except for the close-in planets, the mean eccentricity is in the range 0.2–0.3, with no evidence for dependence of eccentricity on orbital size. The most favored explanation, mentioned by Marcy, Lin, and others, is to somehow pump the eccentricities once the gas has gone. Probably the only hope for this is to invoke some kind of non-linear dynamics or long-term chaos which can come take effect on a timescale much longer than that required for planet formation and disk dispersal.

### 4.3. *Discriminatory comments*

Can we decide between the Belloc and *Python* approaches to planetary formation? Or does the truth lie somewhere in between? Until the discovery of Earth-like planets elsewhere, or the finding of high metal abundance in some observed gas giant, there is, as yet, no direct evidence for the occurrence of core accretion. But there *are* two sets of observations which could be the key to discriminating between the two.

First there is the finding, reported here by Jeff Valenti, that the presence of a planetary system around a star is strongly dependent on the metal abundance of that star, with the probability varying approximately as the square power of the metallicity. Since the core accretion model depends entirely on the presence of metals, this is really *sine qua non* for that model. But, as we have seen, the gravitational instability model (if it works at all) depends crucially on cooling processes in the disk, which are also likely to depend strongly on heavy element abundance. Thus, both models can probably be brought into line with this result, but as yet the details have yet to be worked out.

Second, there are the dust/debris disks reported on by Mike Meyer and Lynne Hillenbrand. The very fact that there *does* seem to be dust and other solid debris in the form of a disk presumably left over from the star formation process indicates that such dust and debris can indeed form, even if the theories as to how it actually does so are not yet complete. The statistics have yet to come in on frequencies and lifetimes, but the very existence of these immediately suggests that at least some parts of the core-accretion scenario seem to ring true. But to be fair, even if one did form many or most of the observed planetary systems through self-gravity, there is no reason to suppose that what is left over might not have formed a debris disk. Indeed, it is the imaging of these disks which might be able to give us clues as to how far from the central star planets can actually form, or end up. Not discussed at the meeting was the body of work (e.g., Wyatt et al. 1999; Quillen 2005; Wyatt 2005) on the observed structure in debris disks which can be interpreted as being caused by Jupiter-mass objects at distances of many tens to hundreds of AU from the central star. If gas giant planets are indeed present at these distances, then that would substantially extend the parameter space of observed planetary systems well beyond those plotted in Figure 1. Ironically, this would also give a headache to the standard core-accretion scenario, because the timescales for forming gas giants at such large radii greatly exceed the dispersal timescales for gaseous disks. Indeed, in answer to a question after his talk, Doug Lin went so far as to claim that if a planet was found that far from the central stars, then his models could not explain it. I am sure that is not true. At these distances, however, as the presentation by Schneider of a $\sim$5 Jupiter-mass object some tens of AU from a $\sim$25 Jupiter-mass brown dwarf shows, we start having to address the question of what is a planet, and what is not.

## 5. Closing remarks

Finally, putting my "Old Fart" cap firmly back on my head, I would like to make a couple of comments on the style of the presentations. First, the ubiquitous use of PowerPoint, and the Internet broadcasting of the talks, now means that presentations are very static. The speaker is now apparently chained to a laptop and now stands like a tailor's dummy prodding it from time to time. This seems to take a lot of the dynamism and drive away from oral presentations. Second, a large number of the speakers seem to have forgotten that it is a good idea when putting up a plot to explain the axes. As usual, Doug Lin was the worst offender. Agreed, he did only show us 23 slides in his 30-minute talk, but each slide contained up to 10 plots, most of which he did not even refer to. My favorite was a color-coded diagram which had the $y$-axis unlabeled, but had the $x$-axis labeled "Burkert." What Andi had done to deserve this I do not know.

As I mentioned before, this is a lively field, and is one which is being driven mainly by observational discoveries. For this reason, I would like to leave the last word with the theorists. The quote is taken verbatim from a response Doug Lin gave in answer to one of the questions at the end of his talk. It is spoken from the heart, on behalf of all theorists, and a sentiment with which I suspect we would all agree: "I am hoping—but I haven't yet completely demonstrated all this."

## REFERENCES

BEER, M., KING, A. R., LIVIO, M., & PRINGLE, J. E. 2004 *MNRAS* **354**, 763.
QUILLEN, A. C. 2006 *ApJ*, **640**, 1078,
WYATT, M. 2005 *A&A* **440**, 937.
WYATT, M., DERMOTT, S. F., TELESCO, C. M., FISHER, R. S., GROGAN, K. HOLMES, E. K., & PIÑA, R. K. 1999 *ApJ*, **527**, 918.